THE UNIVERSE AND ITS ORIGINS

The
UNIVERSE
AND
ITS
ORIGINS

From
Ancient Myth to
Present Reality
and Fantasy

EDITED BY

S. Fred Singer

AN ICUS BOOK
PARAGON HOUSE
NEW YORK

First edition, 1990

Published in the United States by
Paragon House
90 Fifth Avenue
New York, NY 10011

An International Conference on the
Unity of the Sciences book

Manufactured in the United States of America

The paper used in this publication meets the minimum requirements of
American National Standard for Information Sciences—Permanence of Paper
for Printed Library Materials, ANSI Z39.48-1984.

Library of Congress Cataloging-in-Publication Data

The Universe and its origins
"An ICUS book"
Includes Index
1. Cosmology. 2. Cosmogony. I. Singer, S. Fred
(Siegfried Fred), 1924–
QB981.U552   1989   523.1   87-20187
ISBN 0-89226-049-1

Contents

——————— Part **X** ———————————————————
Epilogue

THE UNIVERSE AND ITS ORIGINS

Facing the Universe

S. Fred Singer

How do human beings view the universe, its structure, its origin, and their role in it? Are there other sentient beings out there, have they visited the earth, and will we visit other worlds? These questions have occupied man for millennia, and probably even before recorded history. Foremost has been the possible influence of celestial bodies on human destiny on the earth. Man's relation to the universe, and to its creator, forms the basis of all religious experience.

Widely different cultures have produced surprisingly similar ideas about the structure and origin of the physical universe. In the introductory paper, anthropologist R. Patai recounts the great variety and essential similarity of ancient myths about cosmology and cosmogony. Shaked presents an analysis of myths and the purposes for which they were created.

But myths have not disappeared; they have merely assumed a modern form. An outstanding example is the Velikovsky controversy that surfaced around 1950. The papers by Motz, Oberg, Stiebing, and Bauer discuss different aspects of Velikovsky's efforts to support historical events, such as the biblical exodus, by incorrect scientific theories.

Astrology is another remnant of ancient myths. Divining the future from the positions of stars and planets became a widespread enterprise in human history, and represents a utilitarian forerunner of astronomy. In spite of its rejection by the science establishment, astrology has survived in one form or another. Horoscopes guide the lives of many people; celestial events, such as eclipses and appearances of comets, still appear as a threat to them, as they did in ancient days. Stehling and Tolbert, both astronomers, discuss the history of astrology and why it survives.

The modern view of cosmology, based on the "big bang" theory, is expounded by Chiu, with technical additions and alternative views by von Hoerner, who also discusses the many problems with the current view. To the nonspecialist, these ideas may appear only slightly less mythical than the ancient cosmogonies; it is difficult to communicate concepts that do not conform to our everyday experience, such as four-dimensional space and exotic subatomic particles (such as quarks).

The origin of the Solar System, about 4.6 billion years ago, is closer to us in both space and time. Lewis presents a current picture that is generally accepted, while Lyttleton points to different concepts.

Comets may be galactic visitors to our Solar System rather than original members. According to S. Clube, a giant comet and its breakup products could have been prominent in the sky at the beginning of recorded history and influenced the development of myths about the universe. But Stiebing cannot find evidence, either written or archaeological, to support Clube's fascinating hypothesis.

Astronomical observations were not only the basis of ancient calendars but provided the impetus for modern science, especially Newtonian and Einsteinian physics. Thanks to advances in observational technology, we now have a good picture of the properties of the universe. Space cameras and space telescopes are giving us an even clearer and more detailed view, especially of the planets in our solar system.

Despite all advances, some fundamental problems remain. Did the universe start with a "big bang," and will it keep expanding forever? How far back in time can we trace its beginnings? What is the nature of quasars and "black holes," and what is the cosmological importance of such objects? Just how did the Solar System form, and how unique is it? Why does the earth have oceans, life, and an oxygen atmosphere? And are these really unique, or could there be populated "earths" in other solar systems? Can and did life develop elsewhere?

Some of the solutions to these problems are within reach, perhaps within the next decades. Space telescopes may discover other solar systems. Exploration of Mars may allow us to carry out a program of "comparative planetology" and solve the life and climate riddles, furthering our understanding of climate change on the earth. Radio searches may find evidence for extraterrestrial intelligence within our galaxy.

Space travel, of course, became reality when man first set foot on the moon in 1969 as part of the Apollo program. Mars beckons beyond the moon—being easy to reach, with many scientific and even economic payoffs. O'Leary espouses a manned mission to the moons of Mars, Phobos, and Deimos, as a first step into the Solar System. Oberg concurs and discusses attractive modes of international cooperation. But not all scientists agree; Singer deals with some of the objections to manned space flight.

Shostak describes the ongoing search for extraterrestrial intelligence, supplemented by a relevant discussion by von Hoerner. But the probability of life beyond the earth depends on many uncertain parameters. This probability could change from near-zero to near-one if evidence for crypto-life, or even fossil life, is ever found on Mars—now that we know that early Mars was both warm and wet.

With modern scientific advances, however, the old myths are back, this time in the form of UFOs and "space visitors," instead of gods, demons, and spirits. In this controversial area, Maccabee marshals evidence for TRUFO's ("true" observations of unidentified flying objects) and suggests that they cannot be accounted for by conventional explanations. Oberg strongly disagrees and Maccabee rebuts. We are then left with the conclusion that there may well be unexplained sightings, but that their eventual explanation may not require extraordinary assumptions about space visitors. Stiebing concentrates on reports about ancient space visitors but shows that the historical and archaeological evidence either does not exist or has been misrepresented.

The present volume thus examines myths about the universe, and about intelligence within it, from ancient times to the present: archaeologists, anthropologists, historians, physicists, astronomers, cosmologists, space engineers, and astronauts have had their say. It remains for Bova, a distinguished space expert and science fiction writer, to fashion an epilogue, to tell us why myths and fantasies are fundamental human needs as basic as the drive for a greater scientific understanding of our physical universe.

PART I

MYTHS
OF THE
UNIVERSE

2

Cosmogony and Cosmology

Raphael Patai

Introduction: The Four-"gonies"

Some twenty-five years ago, while studying biblical myths and comparing them with the myths of other ancient Near Eastern peoples, I formulated a definition of myth that read: "Myth is a traditional religious charter, which operates by validating laws, customs, rites, institutions, and belief; or explaining socio-cultural situations and natural phenomena; and taking the form of stories, believed to be true, about divine beings and heroes."[1]

If this definition is valid, as I still believe it is, it must hold true also for that specific category of myth that deals with questions relating to the origin, structure, and functioning of the universe. In effect, myths of the cosmos *are* attempts at the explanation of the grandest and most awe-inspiring natural phenomena observed by man and impinging on his consciousness. They do take the form of stories believed to be true about divine beings and heroes—who in this context occasionally have animal forms—and they answer the questions that arose in the mind of man as soon as he began to contemplate the world around him. These include, first of all, the question of how the world and its constituent parts came into being, and they go on from there to ask: What is the nature, the true character, of the visible parts of the world? How is the world structured? What is the relationship between its various parts? Why and how do these parts move?

The answers to the first of these questions are usually termed *cosmogony*, a term first used by Plutarch (ca. 46–120 c.e.), which can be defined as theories accounting for the origin of the universe, and

7

describing in mythical, philosophical, or theological terms the events to which the cosmos owes its existence. The answers to the rest of the questions mentioned are supplied by *cosmology*. This term was introduced as late as 1730 by Christian Wolff (1679–1754), the famous German philosopher and mathematician, best known for his systematization of scholastic philosophy. A few decades later, Kant (1724–1804) introduced the term *Weltanschauung*, usually translated as "worldview," as a synonym of cosmology. Cosmology comprises theories of the structure and nature of the universe, although it often includes theories pertaining to its origin.

A closer look at the cosmogonical and cosmological myths discloses that they can be arranged into a number of distinct types or categories, whose consideration as such can be helpful in understanding the points of view from which prescientific man approached the phenomena of the world surrounding him. One of these types of cosmogonical and cosmological myths left its traces in the Bible, and the biblical view of the origin and structure of the universe, in turn, became the foundation of the worldview of Judaism, Christianity, and Islam—that is, the majority of mankind. In Christian Europe, it remained dominant until the Renaissance, when it was gradually modified and/or replaced by successively broader vistas, while in the Muslim world it continued to hold its own until the nineteenth century, when Westernization began to chip away at it.

In the early traditions of the Near East, Greece, and other parts of the world, no distinction was made between cosmogony and theogony, that is, between myths relating to the emergence of the cosmos on the one hand, and to the birth of the gods on the other. To complicate matters, the most famous example of the combination of these two subjects, the one contained in the *Theogony* of Hesiod (eighth–seventh centuries B.C.E.), also bristles with stories about the origins of highly sophisticated abstractions that are personalized: Chaos, Hesiod says, "came into being" before all things, and bore Erebos (Darkness), Night, Eros, and Desire. From the union of Erebos and Night sprang Doom, Continence, Discord, Misery, Vexation, Joy, Friendship, Pity, etc. Other abstractions, such as Terror, Anger, Strife, Intemperance, Oblivion, Fear, and Pride, are said to have resulted from unions between Air and Mother Earth.[2]

In addition, there is in early Greek myth a feature that leads straight into archaic astronomy: it consists of stories telling of the transformation of gods, heroes, humans, animals, and monsters into stars or constellations. These myths supply answers given by early speculation to questions about the origin of features observed in the star-studded nocturnal sky, thus creating a subdivision of cosmogony that can be termed *astrogony*—dealing, in mythical form, with the origin of stars and constellations. And there is, of course, yet another type of myth telling about the origin of man, so that *anthropogony* together with theogony, cosmogony, and astrogony form the four basic -gonies, the inseparable mythical quartet of speculations about origins, which for thousands of years satisfied man's desire to know what was, and how that which is came into being.

In the Beginning There Was Not . . .

It is characteristic of the limited nature of human imagination that the great question of how the universe came into being was, as a rule, answered by an evasion: the answer usually given was that the world as known to the observer was the result of a transformation from a preceding, different state; but the question of how that earlier condition originated was, as a rule, not raised. Moreover, most cosmogonies avoid saying anything positive about the pre-cosmic state of affairs, and instead take refuge in a series of negative statements whose purport is merely to emphasize the contrast between the cosmos and the chaos that preceded it. They describe the chaos by listing a number of salient features of the cosmos as *not* having existed. A famous example of this type of solution to the problem of how to characterize the otherness of chaos while saying nothing positive about it, is the *Enuma Elish*, the Akkadian creation epic (dating from the first half of the second millennium B.C.E.), which begins:

> When on high the heaven had not been named,
> Firm ground below had not been called by name,
>
> . . . .
> No reed hut had been matted, no marsh land had appeared,
> When no gods whatever had been brought into being,
> Uncalled by name, their destinies undetermined—
> Then it was that the gods formed. . . .[3]

A much older bilingual poem (in Sumerian and Babylonian) presents the negative features of the pre-creation world in greater detail:

> No holy house, no house of the gods in a holy place had as yet been
>   built,
> No reed had grown, no tree been planted,
> No bricks been made, no brick-mould formed,
> No house been built, no city founded,
> No city built, no man *(Adam)* made to stand upright,
> The deep was uncreated, Eridu unbuilt,
> The seat of its holy house, the house of the gods, unerected,
> All the earth was sea. . . .

Then the poem goes on to tell how the familiar world, the cosmos, emerged:

[Ea] tied [reeds] together to form a weir in the water,
He made dust and mixed it with the reeds of the weir,
That the gods may dwell in the seat of their well-being;
The cattle of the field, the living creatures in the field, he created
The Tigris and Euphrates he made and met them in their place,
Giving them good names.
Moss and sea-plant of the marsh, rush and reed he created.
He created the green herb of the field,
The earth, the marsh, the jungle,
The cow and its young, the calf, the sheep and its young, the lamb of
    the fold. . . .[4]

It is this type of negative approach that is reechoed in Genesis 2, where we read that before God began his creative work, "No shrub of the field was yet in the earth, and no herb of the field had yet sprung up; for the Lord God had not caused it to rain upon the earth, and there was not a man to till the ground" (Genesis 2:5). Compare with this the statement in the Rig-Veda to the effect that in the pre-creation state of the world there was "neither non-existence nor existence, there was no air, nor sky that is beyond it. . . . of neither night nor day was any token. . . ."[5] The oldest of this type of cosmogonies comes from ancient Egypt where it was believed that land and water evolved in the primal chaos of the universal ocean (Nu or Nun), when "not yet was the heaven, not yet the earth, men were not, not yet born were the gods, not yet was death. . . ."[6] The same idea is expressed with masterly brevity in the Upanishads: "In the beginning this [universe] was non-existent. It became existent."[7] Likewise a Polynesian creation myth states that in the beginning "There was no sun, no moon, no land, no mountain, all was in a confluent state. There was no man, no beast, no fowl, no dog, no living thing, no sea, and no fresh water."[8] According to another version of the Polynesian creation myth, in the beginning "there was no earth, there was no sky there was no sea, there was no man."[9] Also in the ancient Teutonic and Scandinavian mythology similar ideas find expression. According to the Völuspá:

There was, in times of old, where Ymir dwelt,
Neither land nor sea, nor gelid waves;
Earth existed not, nor heaven above,
There was a chaotic chasm,
And verdure nowhere. . . .[10]

The Primal State

However, there are also numerous myths about the origin of the universe that are not satisfied with this type of minimal or negative cosmogony, and try to supply instead at least a few generalized features attributed to the age that preceded the coming into being of the cosmos, and, taking them as the starting point, proceed to tell how the cosmos originated, how it is constituted, and how it functions at present. The best known sub-variety of these mythical cosmogonies is that of the creation myths. Creation myths are those cosmogonic myths that explain the origin of the universe by attributing it to the creative work or words of one or more deities, or to some other extra-human beings.

Broadly speaking, the myths of creation fall into two categories: the much more widespread type speaks of the creation of the universe of the preexisting beings or materials, so that they are, in effect, not myths of creation but myths of transformation (metamorphosis); while the much rarer type tells about *creatio ex nihilo*. Of this latter, the biblical creation myth of Genesis 1 is the best known, but by no means perfect, example. The tenor of Genesis is that of creation out of nothing, but the text says nowhere explicitly that God created the world *ex nihilo*. That doctrine had to wait several centuries until it was finally recorded in the second century B.C.E in 2 Maccabees, which states that God created heaven, the earth, and all that is in them, including the seed of man, out of nothing.[11]

The most frequently encountered type of cosmogonic myth is the one which represents the primeval chaos as space filled with water, or, to put it differently, considers the primordial water the substance of the uncreated chaos. Certain psychological schools, of course, have a field day with the concept of the original chaos as water, especially when it appears, as it frequently does, together with the cosmic egg image (see below), since the mythical primordial water lends itself to being interpreted as a reflection of the amniotic fluid that surrounds the fetus in its mother's womb. Since there can be no doubt that both cosmogonic and cosmological myths are in most cases projections on a vast scale of that which man knows from his immediate environment, the parallel between amniotic fluid in the human microcosm and the watery chaos in the macrocosm quite naturally suggests itself. And since a chaos that consists of nothing but water must of necessity be devoid of light, next to wateriness darkness is the most common feature attributed to the primordial chaotic state.

Darkness and water, then, are the descriptions given of chaos in numerous primitive cosmogonies in many parts of the world, as well as in those of many ancient civilizations. The Akkadian creation epic, whose opening words were quoted above, continues by stating that prior to the beginning of the creation process nothing existed but "primordial Apsu the begetter" of everything, who is a mythical personification of the fresh waters, and "Mummu [mother] Tiamat," the personification of the sea, "who bore them all, their waters commingling as a single body."[12] That is to say, the pre-creation chaos was a condition in which nothing existed except an undifferentiated body of waters in which sweet and salt waters were commingled.

A critical reading of Genesis 1:1–2 discloses that despite its author's unmistakable intention of portraying the origin of the universe as a *creatio ex nihilo* performed by the all-powerful and only God *(Elohim)*, traces of a mythical tradition postulating the existence of a primordial chaos had survived in ancient Israel: the text refers to the uncreated presence of *Tohu vaBohu* (chaos), *Hoshekh* (darkness), *Tehom* (the deep, a term related to Tiamat), and *Mayim* (waters). This was, so to speak, the raw material that stood at God's disposal when he embarked upon his six-day labors of creation.[13]

The Brahman creation myth, contained in the Satapatha-Brahmana, opens with the statement, "Verily, in the beginning this [universe] was water, nothing but a sea of water."[14] According to the Rig-Veda, in the pre-creation state of the world "darkness there was at first. . . . this all was water. . . ."[15] In the creation myth of the ancient Mayas as recorded in their sacred book, the Popol Vuh, in the beginning all was calm, silent and motionless, and nothing existed save the empty sky and the calm water, the placid sea, alone and tranquil, and there was only immobility, and silence in the darkness, the night.[16] Likewise in Maori cosmogony, in the beginning "the universe was in darkness, with water everywhere, there was no glimmer of dawn, no clearness, no light. . . ."[17]

These few examples will have to suffice to illustrate the most widespread cosmogonic myth that tells about the emergence of the cosmos from a primordial dark and watery chaos.

The Cosmic Egg

Another familiar image used by archaic man in his cosmogonic speculation is that of the egg. The egg is a small-scale mystery upon which archaic man pondered in many places: how can the egg, which contains nothing but two undifferentiated jelly-like liquids, produce a few weeks later a bird or a reptile complete with all of the many parts of its body? The answer, of course, eluded him, but the potential of cosmic meaning contained in the egg did not. Hence cosmogonic myths from such widely scattered places as Egypt, India, Greece, Finland, Estonia, Borneo, Tahiti, the Society Islands, Hawaii, New Zealand, and Africa, assert that the pre-creation shape of the primordium was that of a gigantic egg.[18]

For the cosmos to be able to emerge, the world egg had to be split in two, as Marduk split Tiamat. Splitting in two often signifies the separation of the undifferentiated continuum of the sky and the earth, or of the Upper Waters and the Lower Waters, so as to produce a space between them in which the creatures can move about. A considerable number of myths belongs to this type, telling of the god or hero who forced apart the chaotically intermingled elements. The most famous of these myths is the ancient Egyptian one which recounts how Shu, the god of the air, raised Nut, the sky-goddess, from Geb, the earth-god.[19] This Egyptian myth of origin is exceptional inasmuch as it makes the sky feminine and the earth masculine—gender attributions that reflect in the first place the fact that in rainless Egypt the life-giving waters, widely considered the equivalent of the male sperm fructifying the earth, originate not from the sky but from the earth, in the form of the River Nile. In other cosmologies the rain-giving sky is masculine and the earth that receives the blessing of the rain is feminine. But common to the Egyptian and many other myths of origin is the notion that the sky had first to be raised up from the bosom of the earth before the cosmic order as we know it today could be established.

Cosmogonic myths explaining in this manner the origin of the sky on high, of the earth beneath, and of life in its myriad forms in between are found in many parts of the world, including Babylonia, Mongolia, India, China, Indonesia, Micronesia, Central and Western Polynesia, Hawaii, Samoa, New Zealand, North and South America, and Africa. They are grouped conveniently by Stith Thompson in his *Motif Index* under the "Raising of the Sky" (A625.2). In fact, they are so widespread that one suspects that they constitute a classical example of the Bastianian *Völkergedanken*.[20] Stripped to its barest essentials, the separation of the sky and the earth is but a variant of the more basic myth that tells about the splitting of the cosmic egg into two as a prerequisite for the processes of creation to be able to begin.

Creation by Deicide-Parricide

A widespread type of cosmogonic myth is the one that accords the chronological primacy to an original generation of monstrous or gigantic deities. These, at a certain point in time, are killed by members of a younger generation of gods, who then proceed to fashion the existing cosmos from the bodies of their colossal divine progenitors. The classical example is found in the *Enuma Elish*, the Akkadian creation epic, already quoted above. In it Marduk kills his ancestress Tiamat, splits her body in two, and makes the vault of heaven of one half, and the earth of the other. Parts of Tiamat's body become geographical features either familiar to the people of Mesopotamia: from her eyes flow the great rivers Tigris and Euphrates—or imagined by them to exist: from a loop of Tiamat's tail Marduk fashions the link between the sky and the earth.[21] Kingu, Tiamat's son and consort, is also killed, and of his blood Ea, the father of Marduk, creates mankind.[22]

The Hittite myth of theogony-cosmogony tells of a succession of several generations of gods, each of whom kills or vanquishes his predecessor. One of them, Kumarbi, bites off and swallows the genitals of his father Anu, becomes pregnant as a result, and gives birth to the River Tigris and to the storm-god who, in due course, defeats Kumarbi.[23]

In contrast to the gory tenor of these Babylonian and Hittite myths of origin, the oldest Egyptian creation myths reflect, not human cruelty, but merely frailty and concupiscence. According to the solar theology of Heliopolis, the god Re-Atum-Khepri, whose three names stand for the sun at noon, the setting sun, and the rising sun, in that order, performed the act of creating the first divine couple, Shu (the air) and Tefnut (his female counterpart) by masturbating or spitting.[24] (Let me interpolate here that creation by masturbation is found also in an Easter Island myth, creation by spitting in Melanesia, and creation by coughing in Mono Alu.)[25] Shu and Tefnut, in turn, gave birth to the god Geb (earth), and the goddess Nut (sky), of whose tight embrace and separation by Shu we have just heard. Once Geb and Nut were separated, and Nut was lifted up high into the sky, Osiris, Isis, and other gods could be born, and Egypt, the kingdom of Horus, could be established.[26] This is but one of several ancient Egyptian cosmogonies; others were associated with Memphis, Hermopolis, and other great Egyptian religio-cultural centers, but limitations of space prevent their presentation.

Greece, in the bloodiness of its mythical cosmogony, is nearer to Babylonia and the Hittites than to Egypt. The Greek myth, too, traces the origins of the world to a series of parricides and bloody scenes in which the protagonists are gods symbolizing or representing primeval elements. Uranos, the sky-god, is castrated and supplanted by his son Kronos, whose name bears a resemblance to, although has no necessary connection with, the Greek term for time, *khronos*. Kronos, in turn, is castrated by his son Zeus, who, as a result of this unfilial act, becomes the king of the gods. From the blood and parts of the severed genitals of Uranos that fell upon the earth were born Aphrodite, the Erynies (furies), the giants, and tree-spirits,[27] as well as certain islands in the Aegean.[28] In Scandinavian mythology (as exemplified in the Völuspa), the gods Odin, Vili, and Vé kill their great-grandfather Ymir and fashion the world by filling the awesome and dark abyss with his huge body.[29]

Generalizing, one can state that polytheistic myths of cosmogony, which are often intrinsically commingled with theogony, are suffused in most cases with streaks of sensuality, callousness, and even cruelty, which no effort at symbolic interpretation can mitigate. The gods of Mesopotamia, Egypt, Syria, Canaan, Greece, and Rome have insatiable sexual appetites; they put to death and devour their offspring, while sons emasculate and kill their fathers, murder and dismember their mothers, and brother does the same to brother. Goddesses, seized with unexplained frenzy, tear to pieces their sons or other persons, and then eat them. Gods inflict all kinds of cruel and unusual punishment on each other and on men whose guilt is but inconclusively established.[30] As against this barbaric company of divinities whose acts give expression to some of the darkest features of man's evil inclinations, the biblical God stands out as a beacon of light, even in his earliest and sternest manifestations, as he revealed himself to the first heroes of Genesis, and even more so in the majestic, compassionate God concept developed by the great Hebrew prophets, the creators of universal ethical monotheism. No wonder that Christianity and Islam, the two triumphant daughter religions of

Judaism, were able between them to convert (even though often not without the use of force) more than half of the human race to their respective versions of monotheistic faith.

It is precisely the intrinsic otherness of biblical religion when viewed against the background of the pagan world that makes it so challenging and intriguing to try to show that traces of polytheistic cosmogonies have survived in the Bible. Such survivals undoubtedly can be found, as in the biblical passages that allude to God's fight with a primeval dragon, or even in the mere fact that, despite the ostensible tenor of the Genesis creation story about the divine fiat that produced the cosmos out of nothing, a number of pre-creation chaotic elements managed, so to speak, to slip into the narrative and can be discerned by modern scholarly acuity. I myself fell under the spell of this challenge and succumbed to the lure of demonstrating the existence of such mythical elements in Genesis, as evidenced by the book *Hebrew Myths*, which I wrote jointly with my friend Robert Graves in the early 1960s. What I did not emphasize in that book, but want to do now, is that those few archaic features notwithstanding, the Genesis account of creation by divine will and word is a unique achievement of the ancient Hebrew religious genius, which places biblical cosmogony in a category *sui generis* in the multiplicity of global creation myths.

Cosmos Out of Man

A number of cosmogonies attribute primacy to man over the cosmos by telling about the fashioning of the world out of the body of a primordial gigantic man, or of some other monstrous but quasi-human creature. The myth of the creation of the world out of the body of man was the subject of an early study by Jacob Grimm, the great German mythologist, who assembled much material to show that this myth preceded its reverse, which tells of the creation of man out of the earth, familiar to all from Genesis 2:7.[31] Other researchers opined that the notion that the world came into being as a result of a primordial human being having been sacrificed is based on the annual rite of sacrificing a man (or an animal) as part of a vernal ritual performed for the fertilization of the earth.[32] A third explanation has it that in archaic view the primal matter of which the world was fashioned had to be a living and thinking being, and only man was such a creature.[33]

The primeval giant, whose body supplied the raw material for the fashioning of the cosmos is occasionally envisaged as an enemy of the gods, and this is why he is killed by them. In this version of the cosmogonic myth the fashioning of the cosmos out of the immense carcass is almost an afterthought and nothing more. This is the impression one gets from the Babylonian myth of Marduk and Tiamat, of which we heard above, and which is retold in detail by Berossus, Babylonian priest and author, of the third century B.C.E.[34]

In a parallel Indian myth, the gods sacrifice Purusa, the first gigantic man, and out of his body fashion the world: "When the gods cut Purusa into pieces, into how many pieces did they cut him? By what names did they call his mouth, his arms, his thighs, and his feet? The Brahman was his mouth, from his arms came into being Rajanya, from his thighs Vaysiya, from his feet Sudra. From his spirit came the moon, from his eye the sun, from his mouth Indra and Agni, from his breath Vayu, from his navel the space of the air, from his head the sky, from his feet the earth, from his ears the four winds. Thus did they fashion the world."[35] Other Indian myths tell of the creation of the world from the bodies of other primal deities: Narayana, Vishnu, and Krishna.[36] A variant is found in the Laws of Manu: the creator, "absorbed in meditation, emitted from his own body the various creatures; he emitted even the waters in the beginning, and in them infused seed."[37]

In Persian cosmogony Gayomard takes the place of the Indian Purusa. Out of the body of this primal man came the metals: "The soil upon which Gayomard died is the gold, and from the other lands in which the dissolution of his limbs took place came the various kinds of metals."[38] Another source gives the names of the eight metals that came from the limbs of Gayomard, seven of them associated with the seven planets.[39] In the Pahlavi *Rivāyat* (tradition) to the *Dātastān I dēnik*, a theological work, it is said that the various creations evolved out of a manlike body. The sky was created from its head, the earth from its feet, the waters from its tears, the plants from its hair, the bull from the right hand, and, finally, fire from the mind.[40]

According to Manichean teachings, the primordial man was a gigantic light creature. He fought Satan, was saved, and was raised among the gods. But he did not remain unhurt: as a result of the attack of Satan, light particles broke off from him, and darkness got control over them. They mixed together with corresponding particles of darkness, and in this manner the world came into being with its twofold nature composed of good and evil.[41] Similar ideas are found in the *Zohar*, the great thirteenth century mystical text of Judaism.[42]

A Chinese myth, found also in countries adjacent to China, reports that once upon a time heaven and earth were inextricably commingled, like a chicken's egg, within which was engendered P'an-ku (perhaps "Coiled-up Antiquity"). After thousands of years P'an-ku increased in size, and thus separated heaven from earth. When P'an-ku died, his left eye became the sun, his right eye the moon, and from his breath came the wind and clouds. His voice turned into thunder, his flesh into fields, his bones and teeth into stones and minerals, his body hair into trees and plants, his marrow into gold and precious stones, his sweat into rain, and the parasites on his body became the first parents of the human race.[43]

We have referred above to the Scandinavian myth of Ymir as an example of the killing of the father-god and the fashioning of the world from his body. In view of the blurred boundary line between primordial god and man, the same myth can also serve as an example of the shaping of the cosmos from the body of a gigantic human.[44]

A similar myth is found among the Gilbert Islanders (on the equator, east of New Guinea). They say that Na-Arean killed his father, Na-Atibu, with the latter's consent, took his right eye and threw it into the eastern sky where it became the sun and took his left eye and threw it against the western sky where it became the moon. He scattered his brain over the sky where it became stars, he sowed his flesh over the water, and it became rocks and stones. Then he took the stones and planted them on the first dry land, the Island of Samoa. And from the bones of Na-Atibu grew "the tree of Samoa," the first ancestor.[45]

Ancient Jewish legends also speak of the *Adam Qadmon*, the primal man, who in contrast to Genesis 1, is portrayed as the first of God's creatures. His body was so huge that it filled the universe, and therefore the angels mistook him for a second deity, until God reduced his size to a mere one thousand cubits. From parts of his body originated heaven and earth and all the rest.[46] This, I believe, is the nearest conjunction ever reached by Talmudic and pagan cosmogonies.

Macrocosm–Microcosm

An overview of the source material and studies dealing with mythical cosmogony and cosmology and with their relationship to the central rituals in archaic, ancient Near Eastern, and primitive societies shows that two major thought processes are involved that proceed in opposite directions: at first from man to the cosmos, then from the cosmos to man.

As far as the myths about the structure of the universe are concerned, the major elements in them are built upon observations made by man in his immediate environment: the cosmos is conceived of as a replica on an immense scale of such familiar natural features as mountains, caves, islands, seas, lakes, springs, as well as trees, animals, humans; or else of such man-made structures as tents, huts, houses, cellars. Selected items from among these environmental features are vastly enlarged in human imagination, projected into space, and by this mental process the baffling observed phenomena of earth, sea, sky, luminaries, and stars are made understandable. Thus the heaven is stated to have windows, to be supported by pillars, by mountains, or by a great tree; the sun and moon are said to be kept in a box, a pot, or a pit; or they are a man and a woman; disks carried between the horns of a bull; or set upon the back of a buffalo. What is involved here is, of course, a projection from the small-scale known to the large-scale unknown, so that the latter, by being viewed as an enlarged replica of the known, becomes itself a known. In this manner the threatening and fearsome quality inherent in the unknown is removed, or at least diminished. Through such processes the cosmos becomes not only reduced in scale in the eyes of the beholder, but also humanized, and therewith man's at-home feeling in the universe is enhanced, and his centrality in the cosmos affirmed.

This *reductio ad humanum* finds its most detailed and meticulous expression in the macrocosm–microcosm analogy, in which correspondences are sought and found between features contained in the universe, the great cosmos, and those of the human body, the small cosmos. To take our first example again from ancient Babylonia, parts of the human body were viewed as corresponding to the planets: the human arm was Mercur, the hand Venus, the eye the sun, the mouth Mars, the head Saturn, etc.[47] Similar concepts were found also in China, as well as in Greece, especially in the Stoic school. According to Plato, man was created in the image of the world and after its pattern.[48] The same idea is expressed in the *Bundahishn*, the most important Pahlavi work on cosmology, which is an exposition of the information as provided by the Pahlavi version of the Avesta,[49] and in which the parallelism between the macrocosm and the microcosm is emphasized.[50]

In the Jewish Midrash the same idea is expressed as follows: "Our sages taught: The creation of the world was like unto the creation of man, for everything the Holy One, blessed be He, created in his world he created in man. The firmament is the head of man, the sun and the moon are the eyes of man, the stars are the hair of man. . . ." Then follows a detailed comparison between each of the twelve signs of the Zodiac and corresponding features discerned in the human body. Thereafter, descending to the sublunar world, the author finds that mountains and hills, fields and deserts, forests and trees, big and small animals, rivers and seas, hot and cold winds, all have their equivalents in the body of man.[51]

Clear traces of these old notions are found in Muslim thinking, and especially in the doctrines of the *Ikhwān al-Safā (The Brethren of Purity)*, the name under which the authors of the tenth century *Rasā'il Ikhwān al-Safā (Epistles of the Brethren of Purity)* conceal their identity. This very influential work is a kind of general compendium of the sciences (including arithmetic, geometry, astronomy, music, logic, "bodily and natural sciences," psychical and intellectual sciences, metaphysical and legal sciences) that was intended to be a manual of instruction for the "Brethren."[52] In it man is pronounced to be the symbol *(ramz)* of the universal existence *(al-wujūd)*, a complete miniature model of the universe, that is, a veritable microcosm. Everything that is in the universe God placed also into man: the body is like the earth, the bones like mountains, the brain like mines, the belly like the sea, the intestines like rivers, the nerves like brooks, the flesh like dust and mud, the bodyhair like plants, etc.[53] Similar ideas recur in the writings of al-Ghazālī (1058–1111), the outstanding Muslim theologian, jurist, and mystic, who says, for example, in his *Kīmiyā al-Sa'āda (The Alchemy of Happiness)* that his intention has been to show that man is a great world, and that the human body is the kingdom of the heart and resembles a great city. Hence "man has truly been termed a microcosm."[54] Also according to the *Rasā'il* the human body is like a city.[55] These speculations exemplify the macrocosm–microcosm equation from the point of view of the medieval Muslim urbanite for whom his city equaled, or at least symbolized, the universe.

It is not difficult to discern in these macrocosm–microcosm equations the reversal of the original thought process that aimed at a reduction of the cosmos to human, and hence humanly comprehensible, dimensions. These equations no longer present man as the basis of understanding the cosmos; on the contrary, they postulate the cosmos as the basis of understanding man. Their aim is to render intelligible—not the cosmos, but man, as the microcosm, the miniature replica of the immense features comprised in the macrocosm. The thought processes underlying these equations take the cosmos as their point of departure, and proceed to attribute to man features, properties, and characteristics that the cosmos was believed to comprise. Therewith the process of familiarization with, and mental domination of, the cosmos was truly completed: the cosmos was not only reduced to a human scale, but man was seen as the equivalent of the cosmos, the reflection of cosmic features, the representative of the cosmos on earth, and hence, without doubt, the center of the universe.

The Cosmic Structure

Ancient cosmologies often depict the universe as a three-storied structure, with heaven above, the earth below, and the subterranean world, usually consisting of water, beneath the earth. This is how the image of the universe can be reconstructed from scattered references found in the Bible.[56] The three-storied structure of the world was a commonplace in the ancient Near Eastern cosmologies.[57] A complementary structural notion is that both the heaven (or sky) and the earth are supported by pillars. This idea is found in Babylonia and Egypt, in the Bible and in Greece, as well as in Siberian, Norse, Eskimo, and Tahitian cosmologies.[58]

Here we again discern early attempts to reduce the mysteries of the universe by explaining them in familiar architectural terms: the heaven is held up above the earth by pillars as is a human habitat or a temple, and in the same manner is the earth lifted up above the subterranean realm. Other mythologies, such as those of the Siberians, the Norse, the Irish, and the North American Indians, speak of the upper and lower worlds as being connected to the earth by rivers or bridges.[59] A larger concept was that of the Sumerians who considered the starry sky an image, not of their houses, but of their country as a whole, traversed by rivers, protected by dikes, and divided by canals.[60]

A widespread type of projection of the earthly familiar to the cosmic unknown for the purpose of reducing the latter to dimensions manageable by human understanding is the image of the world as a tree. The world-tree is either the imago of the cosmos as a whole, or is considered the center of the world and the support of the universe. Such images are found both in the ancient Near East (Kishkanu) and in Germanic religion (Yggdrasil). In the Upanishads the eternal, inverted Ashvatta tree, with its roots on top and its branches below, signifies the pure Brahman, the non-death, in which rest all the worlds. Its branches are the ether, the air, the fire, the water, and so forth.[61] In Altaic religions, among the arctic peoples, and in the Pacific area, the world-tree is represented ritually and is connected with the cultic vessels.[62] It appears also in Babylonian, Norse, Irish, and North American Indian mythologies.[63]

Thus we have the cosmos perceived in the shape of either a man-made three-storied structure, or the largest feature found in the vegetative world, the tree.

Temple and Cosmos: Structure

In ancient Near Eastern traditions and in other ancient cultures as well, the Temple represents the universe. In India, Persia, Babylonia, Egypt, Rome, Byzantium, and elsewhere, the most important temples or sanctuaries were considered replicas in miniature of the world.[64] The same idea with reference to Jerusalem Temple is alluded to in Psalm 78:69, which says that the Lord "built His sanctuary like the heights, like the earth which He hath founded forever." That the psalmist speaks of God and not of Solomon as the builder of the Temple is consonant with the biblical tradition according to which David received the "pattern" of the Temple "by the spirit," that is, by divine inspiration (cf. 1 Chron. 28:11–12). In Talmudic tradition it was a fixed tenet that the Jerusalem Temple, down to minute detail, was an architectural model of the universe.[65] The same notion is presented by Josephus Flavius, the first century C.E. Jewish historian, who says in his *Antiquities of the Jews* that the three parts of the Temple—the Court, the Holy House, and the Holy of Holies—correspond to the three parts of the universe: the sea, the earth, and the heavens.[66] Similarly according to Philo of Alexandria (ca. 20 B.C.E.–ca. 50 C.E.), who lived a generation before Josephus, there was a symbolic connection between the universe and the Temple; thus, e.g., the seven-branched candelabrum was a symbol of heaven, and the altar of incense symbolic of earthly things.[67] Likewise, the Talmudic sages considered the seven lights in the great Menorah to represent the seven planets; the "bronze sea"—the cosmic ocean; the twelve oxen upon which that big basin stood—the twelve signs of the Zodiac. The very rock upon which the Holy of Holies stood, the so-called "Foundation Stone," was believed to be the navel of the earth and the first created thing, under which mysterious shafts led down into the depths of the primordial abyss.[68]

The most succinct and striking expression of the symbolic unity between the world and the sanctuary comes from ancient Rome. The Latin term *mundus*, which means the world, the universe, the cosmos, was also the name of the sacred pit in Rome, in the Comitium, the place of assembly near the Forum. It was kept covered year round by the *lapis manalis*, the stone of the *manes*, the shades of the dead, but was opened three times a year to receive fruit offerings thrown into it. The designation of both the cosmos and the sacred pit by the same name suggests the identification of the two by the ancient Romans.

Another variety of the temple-cosmos equation is the notion that the earthly temple is a replica, not of the cosmos, but of a heavenly temple prototype. This idea, too, is found in Babylonia and Jerusalem.[69] A third variety of the same concept is that the earthly city is built after the patterns of a heavenly city. Well known is the notion of a "heavenly Jerusalem" of which the earthly Jerusalem was but a pale copy.[70] In a fourth variety the entire cosmos or certain parts of it are models for town planning, and the actual building of the city is executed in a manner which presupposes the interpretation of the starry sky as a mysterious map that the priestly astronomer-architect knows how to follow.[71] Thus the Babylonian cities had their archetypes in the constellations: Sippar in Cancer, Nineveh in the Great Bear, Eridu in Vela, Babylon in Ceta-Aries.[72] A modern example is that of the Dogon in western Mali who lay out their villages, fields, and houses in a pattern that is in keeping with the creation myth, and, moreover, expressly state that their kinship system, too, is based on that myth.[73]

Although ostensibly notions such as these explain nearby physical structures (such as temples, cities, canals, divisions of fields) or social systems in cosmological terms, what these traditions and beliefs actually achieve psychologically is to diminish the distance between cosmos and man, and between what man sees in the cosmos and what he builds for himself on earth.

Temple and Cosmos: Function

These observations lead us from structure to function. Again, a few examples will have to suffice to illustrate a phenomenon widespread in both time and space. In Babylonia, where the temple was the *imago mundi*, its construction was considered a reiteration of cosmogony.[74] In ancient Israel, where the Temple of Jerusalem was a representation of the cosmos, its building, commenced (according to legend) by King David, was a repetition of creation, and so was the complex ritual of the "Joy of the House of Water Drawing" performed every autumn in the Temple.[75] In addition, the New Year, celebrated two weeks earlier than that feast, was considered a reenactment of creation. As a Talmudic prayer puts it, "This day is the beginning of thy works, a remembrance of the First Day."[76] This prayer is to this day part of the Jewish New Year observance in an almost identical form: "Today is the birth of the world."

In the Brahmanic ritual the construction of the sacrificial altar is considered a "creation of the world."[77]

The common element in these examples, which could easily be multiplied, is that they express the need to repeat—reenact—creation. As Mircea Eliade put it, the "repetition, by actualizing the mythical moment when the archetypal gesture was revealed, constantly maintains the world in the same auroral instant of the beginning."[78] I would add that archaic societies manifest not only the need to repeat annually the great moment of creation, but feel compelled to do so in order to invigorate nature, man, and the entire terrestrial and extraterrestrial universe, through a symbolic repetition of the cosmogonic process that inaugurated the present state of the cosmos, and made the world habitable for man.

It was inevitable that the temple, the sacred center, the representation of the cosmos on earth, should play the central role in these rituals. The temple ritual had, in the first place, cosmic significance— to ensure the well-being of man. As Hocart put it, "The object of the ritual is to make the macrocosm abound in the objects of man's desire."[79]

Of the many rituals performed in the temple with this purpose in mind none was more complex and believed to be more effective than that of the New Year. The typical New Year ritual consisted of detailed and symbolic re-presentation of the events of creation, achieving thereby the reinvigoration of the world and the assurance of the orderly functioning of the cosmic forces upon which depended the well-being and the very existence of man.[80] Therewith the relationship between the cosmos and man was no longer a one-way street. Instead, a reciprocity was established between them. No longer did only man depend on the cosmic powers that be, but also the latter were now recognized to be dependent on man. The sun and the moon, the stars, the forces that animated the earth, and all that existed on it were, it was now known, able to function in the proper manner that was prescribed for them in those great days when chaos gave way to cosmic order, only if man performed that great annual ritual, reenacting the awesome drama of creation and thus rejuvenating and reinvigorating the cosmos and assuring that everything would go on as it should for yet another year.

But much more than the mere reenactment of creation in a temple ritual was involved in the relationship between man and the cosmos as played out on the archaic scene. The forces of chaos, although defeated by the creator-god or gods in the cosmic struggle or the theomachia that had to be waged before creation could be accomplished, were not rendered totally harmless. They continued to exist, albeit bound, subdued, and imprisoned in their subterranean realm, and were always ready to break out and threaten the world with a cataclysm, or total annihilation. The great flood, which destroyed mankind and the world in the days of Utnapishtim, or Noah, or Deucalion and Pyrrha, was an event in which the forces of the deep, or Tehom-Tiamat, of chaos, were able to gain the upper hand. They were believed to continue crouching beneath the foundations of the temples of Hierapolis in Syria, Jerusalem, and other ancient sanctuaries, and their lesser manifestations were seen in the uncontrollable natural catastrophes that from time to time devastated the cultivated lands. Thus in ancient Sumer, for example, where the cultivated land had to be constantly protected from the marshes, floods, and sandstorms representing the forces of chaos, this labor of maintaining order in the face of the constant threat of disorder was considered an activity duplicating the primeval creative work of the gods. In this manner the everyday life of the people was elevated into the realm of divine creativity, while the creative acts of the gods, who first introduced order into chaos, were seen as something akin to the work of man who subdued and kept at bay the destructive forces of untamed nature.

Conclusion

Our cursory review of the meaning of mythical cosmogony and cosmology has, I believe, shown that those speculations and traditions were for archaic man, primitive man, and historic man until the modern scientific revolution, much more than mere answers satisfying his curiosity about the great riddles of the universe. They were charters containing eternal instructions about the performance and meaning of his most important rituals that secured the universe for him and his place in the universe, and for embarking upon great civilizatory ventures such as bringing land under cultivation and building cities. Above all, the myths of the cosmos provided psychological reassurance for man in his relationship to the cosmos by convincingly showing the cosmos to be anthropocentric.

The role mythical cosmogony and cosmology played in the religious, social, and cultural life of the peoples of the ancient Near East and the classical world was first discussed fifty years ago by E. Burrows, and was subsequently taken up by several other scholars.[81] What has, as far as I know, not been so far the subject of scholarly scrutiny is the relationship between mythical cosmogonies and cosmologies and the modern scientific study of the universe. Scholars who will direct their minds to this subject will, of course, find that mythical cosmogonies and cosmologies have historical primacy over scientific observations and theories, a primacy on the order of thousands of years. But those myths did much more than merely precede in time the great science of astronomy. They satisfied the awakening human curiosity about the awe-inspiring visible cosmos. They interpreted the cosmos anthropomorphically, and represented it as anthropocentric, giving man a measure of self-assurance during the long millennia when he was helpless against the overwhelmingly powerful forces of nature. They provided man with what was believed to be practical and effective means of controlling nature, or at least influencing it to his advantage, and of dealing with the divine powers postulated as the moving forces behind and above the visible universe. Above all, by focusing human attention on the cosmos for thousands of years, the mythical cosmogonies and cosmologies paved the way for a gradually sharpening observation of the universe and an understanding of its nature, its structure, and its functioning. Such observation and understanding ultimately resulted in the birth of modern cosmology, astronomy, astrophysics, and space science.

NOTES

1. Cf. R. Patai, "What Is Hebrew Mythology?" *Transactions of the New York Academy of Sciences*, Series II, vol. 27 (Nov. 1964): 73.
2. H. Rose, *A Handbook of Greek Mythology* (London: Methuen, 1933), 19; R. Graves, *The Greek Myths* (Baltimore: Penguin Books, 1955), 1:33.
3. J. Pritchard, editor, *Ancient Near Eastern Texts Relating to the Old Testament*, Second Edition (Princeton: Princeton University Press, 1955), 60–61.
4. *Encyclopaedia of Religion and Ethics* 4:129, quoting Pinchas, *Journal of the Royal Asiatic Society* (1891): 393–408.
5. Rig Veda X, 129, 1–3. Cf. S. Radhakrishnan and C. Moore, editor, *A Source Book in Indian Philosophy* (Princeton: Princeton University Press, 1957), 23, as quoted by C. Long, *Alpha: The Myths of Creation* (New York: George Braziller, 1963), 169.
6. *Encyclopaedia of Religion and Ethics* 4:144, quoting *Pyramid of Pepys* 1, 1, 663.
7. S. Nikhilananda, *The Upanishads* (New York: Harper and Brothers, 1959), 4:218, as quoted by Long, ibid., 134.
8. T. Henry, *Ancient Tahiti*. Bernice P. Bishop Museum Bulletin 48 (Honolulu: Bishop Museum Press, 1928), 339, as quoted by Long, ibid., 141.
9. A Craighill Handy, *Polynesian Religion*. Bishop Museum Bulletin 34, 1927, 11, as quoted by Long, ibid., 122.
10. *Encyclopaedia of Religion and Ethics*, 4:177.
11. 2 Macc. 7:28.
12. Pritchard, loc. cit.
13. Cf. R. Graves and R. Patai, *Hebrew Myths* (New York: Doubleday, 1964), 21ff.
14. J. Eggeling (translated), Satapatha Brahmana XI, 1, 6, in *Sacred Books of the East* (Oxford: Clarendon Press, 1900), 44:12, as quoted by Long, ibid., 130.
15. See note 5.
16. A. Recinos, *Popol Vuh: The Sacred Book of the Ancient Quiche Maya* (Norman: University of Oklahoma Press, 1950), 81ff., as quoted in Long, 170.
17. H. Hongi (translated), "A Maori Cosmogony," *The Journal of the Polynesian Society* 16:63 (Wellington: Polynesian Society, September 1907), 113, as quoted by Long, ibid., 172.
18. S. Thompson, *Motif Index of Folk Literature* A641, A655, A701.1. Cf. Long, ibid., 109–45; M. Eliade, *A History of Religious Ideas, 1. From the Stone Age to the Eleusinian Mysteries* (Chicago: The University of Chicago Press, 1978), 88.
19. R. Anthes, "Mythology in Ancient Egypt," Edited by S. Kramer in *Mythologies of the Ancient World* (New York: Doubleday Anchor Books, 1961), 36–37.
20. A Bastian, *Der Volbergedanke* (Berlin, 1881).
21. Cf. Eliade, *A History*, 70–72.
22. Pritchard, 68.
23. Pritchard, 120–21.
24. Eliade, *A History*, 88.
25. S. Thompson, *Motif Index*, A615.1, A618.1, A618.2.
27. Rose, *Greek Mythology*, 22.
28. Graves-Patai, *Hebrew Myths*, chap. 21.4.
29. *Encyclopaedia of Religion and Ethics* 4:178.
30. Examples can easily be found in textbooks dealing with mythology or history of religions.
31. J. Grimm, *Deutsche Mythologie* (Graz: Akademische Druckund Verlagsanstalt, 1953 [reprint]), 1:471.

32. A. Hocart, *Kingship* (Oxford: Oxford University Press, 1927), 192f.; W. Bousset, *Hauptprobleme des Gnosis* (Gottingen: Vandenhoeck and Ruprecht, 1907), 209–11. Cf. R. Patai, *Man and Earth in Hebrew Custom, Belief and Legend* (in Hebrew) (Jerusalem: Hebrew University Press, 1942), 1:177.

33. V. Aptowitzer, "Arabisch-judische Schopfungstheorieen," *Hebrew Union College Annual*, 6, no. 49, (1929):222; Bousset, ibid., no. 74, pp. 212, 215, 230.

34. H. Winckler, *Die Babylonische Weltschopfung* (Leipzig: Hinrich, 1906), 21; S. Langdon, *The Mythology of All Races, Volume V, Semitic* (Boston: Archaeological Institute of America, Marshall Jones Company, 1931), 290ff.; Hocart, loc. cit.

35. Rig Veda X. 90, as quoted by Bousset, ibid., 210. Cf. Chantepie de Aptowitzer, ibid., no. 22, 215; Hocart, loc. cit.; Eliade, *A History*, p. 224, gives a somewhat different translation, and does not quote the full list of features fashioned from the body of Purusa.

36. Bousset, ibid., 214f.

37. A. Hocart, *Kings and Councillors* (reprint) (Chicago: University of Chicago Press, 1970), 65–66, quoting Laws of Manu 1:8.

38. Bousset, ibid., 202f., 206.

39. Aptowitzer, ibid., 230; Bousset, ibid., 206–7.

40. M. Dresden, "Mythology of Ancient Iran," in Kramer, *Mythologies*, 339.

41. Bousset, ibid., 178f., 181; Oskar Dahnhardt, *Natursagen*, Vol. 1, *Sagen zum Alten Testament*, Leipzig-Berlin, 1907, 25; Scheftelowitz, *Archivfur Religionsgeschichte 28*, 1930, 212ff.

42. E.g., Zohar, Genesis, 23b.

43. Chantepie de la Saussaye, ibid., 1:279; Hilda Arthurs Strong, *A Sketch of Chinese Arts and Crafts*, Peking, 1926, 3; Derk Bodde, "Myths of Ancient China," in Kramer, *Mythologies*, 382–83.

44. Edda 2, 3. Cf. Grimm, *Deutsche Mythologie* 1:440, 464f.; Bortzler, *Archiv für Religionswissenschaft* 33, 1936, 230ff.; Chantepie 1:107; *Enc. of Rel. and Ethics* 5:128; Bousset, ibid., no. 1, 211; Hocart, ibid., 194.

45. Hocart, ibid., 194–95.

46. Cf. Patai, *Man and Earth* 1:180–88, and especially 184. Myths of many peoples about the creation of the universe from parts of man's or a creator's body are catalogued in S. Thompson, *Motif Index* A614.

47. H. Winckler, *Die babylonische Kultur*, Leipzig: Henrich, 1902, 23.

48. Plato, *Timaeus* 44–47.

49. Dresden, ibid., 335.

50. M. Mole, *Culte, mythe et cosmologie dans l'Iran ancien: Le probleme zoroastrien et la tradition mazdeenne* (Paris: Presses Universitaires de France, 1936, Annales de Musee Guimet), vol. 69, 114.

51. Aggadat Olam Qatan, in Adolf Jellinek, *Beth Hamidrash* 5:57–59; cf. Patai, *Man and Earth* 1:166ff.

52. *Encyclopaedia of Islam*, Second Edition, s.v. *Ihkwn al-Safa* by Y. Marquet.

53. S. Nasr, *An Introduction to Islamic Cosmological Doctrines* (Cambridge: The Belknap Press of Harvard University Press, 1964), 96, 99, 101–2.

54. Al-Ghazali, *Alchemy of Happiness*, Albany: 1873, pp. 19, 37, and Ghazzali, *The Alchemy of Happiness*. (Translated from the Hindustani by C. Field), (London: John Murray, 1910), 28.

55. Cf. Masr, ibid., 99.

56. E.g., Gen. 7:11, and especially Ex. 20:4, where the component parts of the world are referred to under three categories: "any thing that is in heaven above, or that is in earth below, or that is in the water under the earth." Cf. also "the deep that croucheth beneath" Gen. 49:25; Deut. 33:13. "The deep" figures repeatedly in biblical poetical imagery, as does the notion that the earth is founded upon the seas or the floods, Ps. 24:2, or that it forms "a circle upon the face of the deep." Prov. 8:27; cf. 28.

57. *Die Religion in Geschichte und Gegenwart*, new edition, s.v. Weltbild, 1C Alter Orient.

58. S. Thompson, *Motif Index*, A665.2; *Die Religion in Geschichte und Gegenwart*, ibid., 6:1615; Job 9:6; 26:11; cf. Ps. 104:3,5.

59. S. Thompson, *Motif Index*, A657–A657.1; cf. A659.3; A661.0.5.

60. *Die Religion*, ibid., 6:1612.

61. Cf. Katha-Upanishad VI.1 and Maitri-Upanishad VI.7, as quoted in *Die Religion*, ibid., 6:1629–30.

62. H. Lubac, "L'arbre cosmique," in Melanges E. Podechard (Lyon, 1945), 191–98; M. Eliade, *Traite d'Histoire des Religions*, Paris, 1949, par. 95ff.; *Die Religion* 6:1630.

63. Cf. S. Thompson, *Motif Index*, A652–652.4.

64. Cf. Patai, *Man and Temple in Ancient Jewish Myth and Ritual*, Second Edition (New York: Ktav, 1967), 105–7; Hocart, ibid., 176, 191; idem, *Kings and Councillors*, Cairo, 1936, 230ff.; Eliade, *A History*, 90, and lit. ibid.

65. Patai, *Man and Temple*, 107–39.

66. J. Flavius, *Antiquities of the Jews* 3:7:7.

67. Philo. *Life of Moses* 3:10.

68. Patai, *Man and Temple*, 31, 86, 108–13.

69. Jeremias, *Babylonisches im Neuen Testament*, 62ff.; Patai, *Man and Temple*, 130–32. Cf. also Louis Ginzberg, *The Legends of the Jews* (Philadelphia: The Jewish Publication Society of America, 1909–1940), Index, 7:256, s.v. Jerusalem, heavenly.

70. Patai, loc. cit., Ginzberg, loc. cit. On the city as the equivalent of the cosmos, cf. Hocart, *Kings and Councillors*, 244–55.

71. Cf. W. Muller, *Die heilige Stadt, Roma Quadrata, Himmlisches Jerusalem, und die Mythe vom Weltnabel* (Stuttgart: ——, 1961), which discusses the cosmic views forming the model for city planning.

72. E. Burrows, "Some Cosmological Patterns in Babylonian Religion." Edited by S. Hooke in *The Labyrinth* (London: Society for Promoting Christian Knowledge, 1935), 60–61.

73. S. Moore, "Descent and Symbolic Filiation." Edited by J. Middleton in *Myth and Cosmos: Readings in Mythology and Symbolism* (Garden City: The Natural History Press, 1967), 70–71.

74. Eliade, *A History*, 61.

75. Patai, *Man and Temple*, 27ff., 58, 84–85.

76. *Babylonian Talmud, Rosh haShana* 27a; Patai, ibid., 69, 68.

77. M. Eliade, *The Myth of the Eternal Return* (Princeton: Princeton University Press, 1954), 78, 80.

78. Eliade, ibid., 90.

79. Hocart, *Kings and Councillors*, 202.

80. Patai, *Man and Temple*, 118–20.

81. E. Burrows, ibid., 45–70; Patai, *Man and Temple* Eliade, *Myth of the Eternal Return*. For recent literature on the subject see K. Bolle, "Cosmology," in the *Encyclopaedia of Religion*, 1986.

3

On the Interpretation of
Cosmogonical Myths

SHAUL SHAKED

Professor Patai's rich paper ends with the observation that there is a parallel between the mental processes underlying myths of origin and mythological cosmologies on the one hand, and those that induce scientific theories concerning the origins and structure of the universe on the other. To supplement his comments, some further points may be made.

The similarity in motivation between science and mythology rests, of course, on the obvious common drive of curiosity. There is, however, more than that involved. There is in that not only a desire to acquire knowledge, which implies a degree of certitude and truth (which science can only supply in a limited sense); there is also a need to take some of the edge off the unknown, an unknown that is inherently not only mysterious but frightening. By bringing about a certain familiarity with the universe, the latter becomes less menacing. Both science and mythology achieve this effect by using analogies borrowed from the familiar environment around or within man. As a result, not only is our comprehension of strange or remote phenomena facilitated and enhanced, but those phenomena themselves become closer and acquire a certain intimacy. The sense of potential danger inherent in objects or movements that have the appearance of a different order of existence is thus neutralized.

By employing with regard to distant and invisible objects the same system of rules that we apply to ourselves and to our close environment, and declaring those rules to be valid for the whole universe, the cosmos is reduced to human-sized space and some of the terror it is apt to inspire is dissipated.

In mythical cosmology the powers governing the universe are

usually represented as humans or as anthropoid divine entities. Even when they are seen as nonhuman, for example, as animals or as natural powers, they are described with human feelings, deliberations, and discourses. In science, in a parallel fashion, we necessarily assume that even in places where direct observation is not possible, even with regard to phenomena that are not within the compass of human accessibility, the same laws apply and the same basic type of matter exists as in our globe. As a characteristic example of the analogical mode of explanation we may refer to the idea that man and the world form two parallel phenomena. The most conspicuous expression of this idea, implicitly present in some mythologies and explicitly in some philosophical systems, is the representation of a structural similarity between the microcosm and the macrocosm (that is, the description of the universe as a large man, a *macranthropos*).

The similarity that exists between some scientific theories of cosmic beginnings and certain mythical motifs where, for example, the former make use of cataclysmic events, or of the notion of primal liquid matter, is perhaps not entirely fortuitous. In each case we have the urge to describe the universe in terms of beginnings as we know them in familiar objects, and apply those notions to a place and time that, by definition, are outside the area in which direct evidence is, or can ever become, practical.

Such comparisons of mythology and science can only be made in very general terms. Even when they are legitimate they should not overshadow the obvious but profound differences between the two modes of activity. Some of the initial impetus for scientific theories may indeed come from an imaginative leap not unlike that of mythmaking, but once a theory is formulated it is subject to verification, questioning, and modification in accordance with observation wherever it is available in conformity with other known laws and theories. It is significant that in the history of science the pioneering theories have often been close to ideas current in religion and mythology. After a series of modifications the current theories may make a full circle and come back to resemble somewhat the initial mythical idea. It would not, however, be quite legitimate to talk, in such a case, of mythology being vindicated.

There are two or three important remarks to be made about mythology, mostly for the benefit of those who have not dealt with this subject professionally before. It would be wrong to regard mythology merely as a series of intellectual statements about the universe and its origins. Even when a myth does seem to be making this kind of statement, it is as much a statement about other things—the individual and his fears, anxieties, and hopes; society and its institutions—as it is about the universe. It was a typical fallacy of earlier days in the study of mythology to regard this field as representing a pre-scientific or pre-philosophical mode of reflection about the cosmos.[1]

Myths are notoriously difficult to define.[2] Patai has provided us, at the beginning of his paper, with a useful and fairly comprehensive definition of myth, one that takes into account practically all its divergent functions and contents. If I wished to improve on it I would think of doing so mainly by suggesting some changes in the order and relative importance of its various elements.

Myth, it may be recalled, was defined by Patai (p. 8) as "a traditional religious charter" that operates by validating and explaining certain things and that takes the form of stories.[3] By defining myth as a charter we assume it primarily to have a quasi-legal function. I believe that this conception would not hold water even with some myths of origins; it would certainly be much harder to justify with most other myths. Much of Greek mythology does not seem to have any charter character at all. It would be difficult, for example, to assign a charter significance to the myths of the birth of Zeus, or to those of Agamemnon, Odysseus, and so on. Even when they do appear to explain and justify an institution or a phenomenon, it is characteristic of myths not to do so deliberately or explicitly. The etiology is left in many cases to be guessed at. I would therefore prefer to place as the primary element of myths not their function as charter, which is rarely present in any conspicuous manner, but their form, that of a story. Myths are basically stories told. These may have as one of their important effects the explanations and justification of existing social institutions and natural phenomena, as well as the cohesion of the social group, but they are not legal documents.

Myth is an expression of a social group. This fact helps to explain how it is able to validate the central elements in the life of the society without necessarily alluding to them explicitly. The very fact that certain myths are being told and retold, even without taking their contents into account, contributes to the sense of cohesion and unity in a society. Ideally, therefore, myths should be studied within their social context, along with the ritual activity, the body of faith, the social structure, and the economic system, although they can also of course be studied on their own from a specific point of view, for example, for their subject matter, as done here.[4]

By analyzing, comparing, and classifying the intellectual responses that myths of different cultures give to the problems of origins,[5] we may indeed gain a better insight as to the type of questions myths of origins tend to cope with and the characteristic forms that their solutions take. We may also detect, as Patai did in his paper, certain themes recurring in several different cultures. We should, however, be aware of a long-standing and extensively discussed problem in scholarly literature—the question whether the occurrence of common themes in different cultures is the effect of diffusion from one area of origin to another. An alternative to this hypothesis is provided by the assumption that the human mind tends to produce similar images and motifs in a spontaneous manner. A typical proponent of the latter theory was C. G. Jung, who coined the term "archetypes" for the notion of recurring images tending to occur, according to him, in an independent manner in various individuals coming from a variety of cultural backgrounds.[6]

In this particular forum we may wish to examine a third hypothesis, perhaps implied in Patai's treatment of the different myths concerning the origins of the world, namely, the hypothesis that the variety of mythical accounts recounting a similar event may be indicative of the fact that the event thus described may have actually taken place within the living memory of mankind. According to the last formulated hypothesis, we may try to find out whether the diverse reports of the same cosmogonic events contain genuine reminiscences of actual historical occurrences. It goes without saying that if the diffusionist explanation or the archetypal theory is accepted as valid, the strength of the evidence for the third possibility, that of a genuine historical reminiscence, is greatly reduced. Under the first hypothesis, the variety of myths dealing with a given theme are no more than so many different variations of a single mythological source, and their number has no bearing on the validity of the report itself. If the second hypothesis is correct—that the human mind tends to create such myths independently and without need for an external stimulus—the more instances we find of a given theme the more we can be sure that it is a characteristic feature of the human imagination.

One may, of course, adopt a nondogmatic attitude and assume that either diffusion or parallel creation provide in different instances the best explanation for the existence of striking similarities in the myths of diverse cultures, and reject the wholesale assertion that all phenomena of myth invariably fall under a single explanation. However, having adopted this point of view, which seems a commonsense one, we may feel that very little room is left for the view that the existence of several similar myths in different cultures constitutes a proof for the contention that certain cosmogonic events took place at a primordial time in the history of the world. They may have actually occurred, of course, but one needs more than a long list of mythological attestations for them to be proven; they have to be supported by evidence from other sources.[7]

As an example, we may mention the wide distribution of the theme of the primordial flood. The many attestations of this myth may be significant only if they can be shown to be independent of each other. If this is the case, they may either point towards an independent reminiscence of a historical event, or they may belong to an archetypal kind of representation. It is a pity that we are not told what conclusion Patai wishes us to draw from the rich material he has brought together. It may safely be stated that there are considerable difficulties associated with each one of the theoretical positions enumerated above.[8] The study of ancient myths is on the whole best left to be carried out within their own narrative structures, with constant reference to the larger social context in which they are embedded.[9]

The idea of creation *ex nihilo*, it may be remarked, is very rarely present in mythology. Even when it seems to be expressed, a closer look often reveals that something did exist from which or by which the creation of the cosmos started. The strict conception of creation out of nothing, although it is often hinted at, is not part of the stock of ideas of archaic mythology. What seems to be a rather prominent feature of such myths, in contrast, is the idea of an order imposed on the primordial chaos.[10]

The discussion up to now has dealt with myths as if they consisted chiefly of their intellectual or concrete factual contents. But an important aspect of myths is that they constitute an imaginative and emotive mode of expression. In this sense, they belong not to the field of science, nor to that of philosophy or history, but to those of art and religion. This observation highlights certain difficulties in the interpretation of myths. The language of myths is that of storytelling, not of intellectual discourse, and as a result we often find ourselves on slippery ground when we try to translate their contents into clear and direct factual statements.[11] Almost any story is capable of more than one interpretation; in fact, it frequently lends itself to several conflicting modes of rendering.

A case in point is the atrocity and immorality (from our point of view) that are manifested in several ancient mythologies. Patai observes that "polytheistic myths of cosmogony . . . are suffused in most cases with streaks of sensuality, callousness, and even cruelty, which no effort at symbolic interpretation can mitigate." I would argue that there is no need to soften or mitigate such cruelty where it exists. One should try to understand it, but this entails admitting that myths, like the human mind, are ambivalent. Does the occurrence of such traits imply approval of such acts as applied to the narrator and his audience? When violence, callousness, or immoderate sensuality are encountered, they are usually the hallmark of the mythical hero—not an ordinary human being—one of the features that set him apart from the rest of mankind. The frequent implication is that violence or licentiousness, tolerated in the mythical context, is strictly forbidden outside it. Could one argue that the inclusion of extraordinary feats of atrocity and lack of moderation serve as oblique expressions of hidden desire to indulge in such acts? This may be so, but then such myths surely serve as release valves, enabling people to resolve some of the pressures and tensions created within them and imposed on them by society. It is also possible to regard them as a way of overcoming terror, a means by which human beings try to be liberated from the dread of assaults by these mighty and awesome powers. It is possible to achieve this liberation by identifying oneself symbolically with those very powers who are allowed to go beyond the limits imposed on mortals.

Viewed from this perspective, one wonders whether a relatively "clean" mythology, such as that of the Bible, can be said to be in any real sense superior to those where excessive doses of sexual and murderous acts are featured. One may indeed inquire whether "clean" mythologies have not lost something that may be essential to the human experience, ending up being in some way poorer for this loss. It is possible, on the other hand, to speculate that this loss is made up by other elements that may satisfy similar needs. Formulating such a question may entail examining the whole field from a different kind of perspective, by trying to apply to it an idea of a harmony between elements, which will probably be impossible to test. The purpose of these remarks is merely to point out that there seems to be little sense in imposing a value-judging attitude on myths, and that it may be better to try and understand them in their own terms.

The motif of the cosmic egg, so effectively described by Professor Patai, may be taken as an example illustrating the multiplicity of meanings capable of being attached to mythical contents. The egg most often signifies wholeness and integrity, but it may also signify lack of distinction, the absence of the definition and contrast essential to the structure of the civilized world. It may further indicate a self-contained entity that needs to have recourse to nothing outside it; but sometimes (as in the Iranian myths of creation) it represents a prison-like trap in which the divine antagonist, the power of evil, is being ensnared.[12] The mythical theme may have wandered from one culture to another, but its sense seems in many cases to have undergone significant changes.

A similar variety of meanings exists in such mythical themes as the primordial chaos, the water, the flood, paradise, and so on. In mythology such a plurality of divergent senses can even be taken to coexist in the same culture on different levels of interpretation. A mythical statement is by its nature ambiguous. It is a story designed to delight and enthrall an audience, and yet it carries a heavy weight of implied meanings—all legitimate—within the range allowed for by the culture to which it belongs.

The thematic approach used by Patai has as its inevitable effect a certain blurring of the sense and structure of individual myths in their own context. One could think of other possible classifications, where a general view of the themes discussed may be retained, while allowing us to distinguish between types of cultures and their respective mythologies. There may be a good reason to separate the discussion of myths emanating from complex and literate societies from those collected by anthropologists outside the main centers of organized culture. Despite certain common traits, the structure and contents of these two sets are quite distinct from each other. Within the first group it seems to make good sense to treat the mythologies of the great polytheistic systems, those of the Ancient Near East, Greece, Rome, and India apart from those of Judaism, Christianity, and Islam, to which the myths of Zoroastrian Iran and of Buddhism may possibly be added. The former group represents a conception of divine power based on divergence and diffusion, while the latter constitutes a group of cultures where a sense of a strictly defined cosmic and divine hierarchy and a strong concentration of power prevail. This contrast in the general structure of the religious civilization has its repercussions in the structure and use of the myths used. Other classifications are certainly also possible; it seems, however, that one should try to avoid treating all human expressions concerning the origins of the world as constituting one interconnected mass of material, and that structural, social, and historical distinctions should be made before the thematic issues are approached.

NOTES

1. The most prominent exponent of this kind of idea was L. Lévy-Bruhl; cf. an analysis of his ideas and a critique in Evans-Pritchard 1965, 78ff. The same kind of attitude is also represented in Lang 1911.
2. See the chapter on problems of definition in Kirk 1974, 13ff.
3. This definition seems influenced by Malinowski 1922, 298ff.; 1926, 19.
4. For a wide range of approaches to the study of myths cf. Kirk 1970 and Kirk 1974.
5. It may be noted that the presentation of four different "-gonies," as done by Patai, may be improved somewhat by talking of only three categories belonging to this group. The study of the origins of the heavenly bodies can hardly be said to constitute an independent province of mythology.
6. An example for the application of this method in the study of myths is provided in Jung and Kerényi 1949.
7. It is particularly inappropriate to quote as supporting evidence myths recounted in sources which are known to be late and dependent on earlier attestations. This applies, for example, to quotations from Manichaeism or the Zohar (see Patai's chapter, p. 00).
8. A very detailed treatment of flood myths is by Woods 1911. The theme is also discussed in various places in Brandon 1963 as well as in Loewenstamm 1962. See also Kirk 1970, 116f.
9. Reducing the study of myths to that of structures and relationships has become fashionable, largely under the influence of C. Lévi-Strauss's work, but one cannot help feeling that certain essential elements tend to be lost in the process. For an anthropologist's critique of Lévi-Strauss see Douglas 1975, 153ff.
10. Cf., e.g., Brandon 1963, 208ff.
11. This, of course, was one of the fallacies of Velikovsky, whose work has been one of the themes of this conference.
12. Some references to the Iranian myth of creation will be found in Shaked 1967 and 1971.

REFERENCES

Brandon, G., 1963. *Creation legends of the Ancient Near East*, London: Hodder and Stoughton.
Douglas, M., 1975. *Implicit meanings*, London, Boston and Henley: Routledge and Kegan Paul.
Evans-Pritchard, E., 1965. *Theories of primitive religion*, Oxford: Clarendon Press.
Jung, C. and Kerényi C., 1949. *Essays on a science of mythology*, New York: Pantheon Books (Bollingen Series, XXII).
Kirk, G., 1970. *Myth: its meaning and functions in ancient and other cultures*, London: Cambridge University Press—Berkeley and Los Angeles: University of California Press.

—— 1974. *The nature of Greek myths*, London: Penguin Books.
Lang, A., 1911. "Mythology", *Encyclopaedia Britannica*, 11th edition, Cambridge, England and New York, 128–144.
Loewenstamm, S., 1962. "Mabbul", *Encyclopaedia Biblica* (in Hebrew), Jerusalem, IV, col. 597–610.
Malinowski, B., 1922. *Argonauts of the western Pacific. An account of native enterprise and adventure in the archipelagoes of Melanesian New Guinea*, London: Routledge and Kegan Paul [Reprinted: 1950].
—— 1926. *Myth in primitive psychology*, New York: W.W. Norton.
Shaked, S., 1967. "Some notes on Ahreman, the Evil Spirit, and his creation", *Studies in mysticism and religion presented to Scholem, G.*, Jerusalem: Magnes Press, 227–254.
—— 1971. "The notions *mēnōg* and *gētīg* in the Pahlavi texts and their relation to eschatology", *Acta Orientalia* 33: 59–107.
Woods, F., 1911. "Deluge", Edited by Hastings J. in, *Encyclopaedia of Religion and Ethics*, Edinburgh, IV: 547-557.

THE VELIKOVSKY AFFAIR

4

How Good Were Velikovsky's Space and Planetary Science Predictions, Really?

JAMES E. OBERG

With a quarter-century of hindsight, we can approach the question of the alleged accuracy of Velikovsky's predictions about space and planetary science. The development and evolution of this claim may provide some insight into the mentality and techniques of argumentation used by Velikovsky, his supporters, and his detractors.

As expounded in a series of books, lectures, and short articles, Velikovsky had created an intricate ballet of celestial encounters that repeatedly devastated the earth (he preferred to call his works a "reconstruction" rather that a "theory"). Venus was one of the key planets, having sprung from Jupiter during a close encounter with an errant Saturn. The newborn world was enveloped in gases and appeared as a comet, swinging deep into the inner Solar System periodically. After a few hundred or a few thousand years, earth's luck ran out; and it nearly collided with comet-Venus. The resulting series of disasters and subsequent near-collisions form the causes of the parting of the Red Sea, the plagues of Egypt, the fall of the walls of Jericho, and the "sun standing still" over Gibeon. . . . such is the "reconstruction" of Velikovsky. Venus, meanwhile, knocked Mars off course while itself settling into a more peaceable orbit; the Red Planet, in turn, made a series of equally devastating visitations seven centuries later.

And all of this constituted only the final act in a drama which Velikovsky envisaged as going back thousands of years. Earlier, he wrote, Mercury, too, had been involved in planetary encounters; Jupiter and Saturn had been double planets, and earth had been a satellite of one of them: Saturn had exploded, showering earth with the waters of the Deluge; the moon had come in from somewhere and had been captured somehow by the Earth. All of this was supposed to have occurred "within the memory of man."

As the years passed, Velikovsky claimed a series of important "successful predictions" about the Solar System, based on his world views: "The space age gave my views a record of confirmations," he wrote in 1977. Earlier, one of his chief disciples was even more specific: "Seldom in the history of science have so many diverse anticipations—the natural fallout from a single central idea—been so quickly substantiated by independent investigation" wrote Ralph Juergens in September 1963. Velikovsky himself treated his theories as "proven," even while he remained a scientific pariah, when he asserted in 1974 that "My work today is no longer heretical. Most of it is incorporated in textbooks, and it does not matter whether credit is properly assigned. . . . None of my critics can erase the magnetosphere, nobody can stop the noises of Jupiter, nobody can cool off Venus, and nobody can change a single sentence in my books."

This attitude is all the more baffling in light of the claims made by Velikovsky's detractors, most notably Carl Sagan. In reply to the assertion that space exploration had verified Velikovsky's theories, Sagan told a San Francisco seminar in 1974 that "To the best of my knowledge, there is not a single astronomical prediction correctly made in *Worlds in Collision* with sufficient precision for it to be more than a vague lucky guess and there are . . . a host of claims made which are demonstrably false." This view was elaborated by NASA space scientist David Morrison in a newsletter debate published in 1979. Wrote Morrison, "Every important prediction he made in 1950 concerning conditions on the planets, such as hydrocarbon clouds on Venus, large amounts of argon in the atmosphere of Mars, recent melting of the lunar surface, large internal heat sources on Venus and perhaps Mars, large-scale recent cratering of earth and moon, and synchronized planet-wide volcanism on earth have been shown decisively to be in error. . . . Every new space mission, such as the recent Pioneer-Venus probes, pounds another nail in the coffin. . . . The cruel truth is not only that astronomical evidence fails to support Velikovsky, but that a great deal that seemed plausible or at least possible when suggested in 1950 has been shown to be incorrect."

It is an interesting exercise to summarize the planetary predictions he made—and later remade in light of what was actually discovered—and compare them to actual space data. Such an effort should restrict itself to such physical measurements and avoid all questions of terrestrial ancient history (which Velikovsky sought to revise chronologically by many hundreds of years) and of interplanetary dynamics (where Velikovsky has called for the actions of immense non-gravitational forces). Instead, by concentrating on the alleged "confirmation" of his planetary predictions, we may now with hindsight be able to determine if his "hits" were significant proofs, coincidences, lucky guesses, or merely Monday-morning-quarterback revisions of originally ambiguous handwaving that could be interpreted ex post facto to mean *anything* (the "Nostradamus Effect").

One of the cornerstones of the edifice of the claimed "successful predictions" of Velikovsky is the heat of Venus. When he wrote *Worlds in Collision* in 1950, the scientific world considered Venus to have surface temperatures only slightly higher than that of earth. Yet space probes later found that the temperatures exceeded 800° F., a figure that seemed to bear out Velikovsky's claim that the newborn planet had at one time been "candescent," or glowing, with excess heat. Later efforts by astronomers to account for the high temperatures by means of a "runaway greenhouse effect" were denounced by Velikovsky as clumsy groping—"completely unsupportable" he called it in 1974, adding that such an idea was "in violation of the Second Law of Thermodynamics."

Actually, the heart of Velikovsky's prediction was that Venus would be giving off more heat than it received from the sun, and that it would still be cooling off. Space probes and recent terrestrial measurements have both provided voluminous evidence *against* these claims, and the Pioneer-Venus data in 1978 showed that the "runaway greenhouse effect" is still very much in the running. One problem with this "effect" is that it depends on trace atmospheric constituents for thermal contributions all out of proportion to their actual mass; so there is still room for doubt.

Velikovsky also claimed that "the presence of hydrocarbon gases and dust in the cloud envelope of Venus would constitute a *crucial test*. . . ." and further that "on the basis of this research, I assume that Venus must be rich in petroleum gases." Neither of these predictions have been borne out by data from American and Russian atmospheric probes. Velikovsky also had suggested that "microbial life able to catalyze can possibly be found in Venus' clouds"—an unlikely development. The surface of Venus will be very plastic, he predicted, with "mountain-high ground tides"—whose absence was established by the radar scans from the Pioneer-Venus Orbiter in 1979 that located solid continents and mountain ranges. He claimed to have predicted the massive atmosphere of Venus, but no records have been found that show that he did so until after it was discovered. Lastly, he claimed that Venus would have "a weak magnetic field," an ambiguous stab that comes up against Pioneer-Venus determinations that there is *no* detectable intrinsic Venusian magnetic field at all (magnetic fields *are* induced by the solar wind, which Velikovsky did *not* predict).

Another voice in the Venus prediction controversy was also raised in mid-1980 when Kendrick Frazier, former editor of the weekly *Science News* and now editor of the quarterly *Skeptical Inquirer*, wrote an analysis of Velikovsky's claims versus recent discoveries about Venus. His article, "The Distortions Continue," was based on extensive interviews with scientists involved with the Pioneer-Venus probes that reached that planet in December 1978, using as a discussion point an advertisement from Velikovsky's publisher that claimed that recent space discoveries had verified many of Velikovsky's predictions.

The scientists' reaction was blunt. "The ad is thoroughly dishonest," wrote one; "The ad, like the book it is promoting, contains more falsehoods in a paragraph than one can refute in a chapter." Dr. David Morrison of the University of Hawaii, a frequent Velikovsky critic, was even more direct: "Several of the statements in the ad are outright lies. I am used to distortions by the Velikovsky supporters, but this ad seems to be particularly reprehensible." Another astronomer put the dispute in perspective, "The Velikovsky hypothesis was *never* controversial among scientists," wrote George Abell. "It is, and was, recognized at once as a crank idea."

The earth's moon played a significant role in Velikovsky's scenarios of planetary encounters. Most prominent among the successful predictions by Velikovsky are the existence of "numerous moonquakes" and of "lunar remanent magnetism," both made in the face of the expectations (or lack of them) of "establishment science."

This is what Velikovsky wrote about the possibility of moonquakes, using the July 21, 1969, issue of the *New York Times* as a forum: "I also maintain that moonquakes must be so numerous that there is a bit of chance that during their few hours on the moon, the astronauts may experience a quake." The astronauts, of course, did *not* "experience" (i.e., "feel") a moonquake, nor did any subsequent astronauts—nor would they if they had stayed for millions of years. As it turns out, moonquakes are very weak and very rare; and it is only because of the extreme sensitivity of Apollo instruments and because of the high seismic conductivity of the super-dry lunar surface that the very tiny moonquakes that occurred there were recorded. But at every comparable level of intensity, the moon is a much, much quieter place than is earth. Velikovsky's description of the moon and of the astronauts' chances of encountering moonquakes were both in error—although a few years later, his followers were making it sound as if he had been right on target: "He said only that quakes would be numerous," wrote Dr. C. J. Ransom in 1975, "and did not suggest that they would be of great magnitude." A skeptic is justified in wondering that if this is the case, how did Velikovsky expect the astronauts (*not* their instruments) to "*experience*" them?

The Velikovsky school refused to concede the point. Wrote Lewis Greenberg, editor of *Kronos*: "With respect to moonquakes, Oberg should consult the October 1971 issue of *Chemistry*. There, he will find a NASA report of 11/20/70 to the effect that equipment left on the moon by Apollo XII recorded an average of one lunar quake a day, with more severe quakes occurring once a month. This hardly sounds 'rare.' . . ."

But it also remains true that these moonquakes are both tiny and rare. I quote from Elbert King's *Space Geology* (1976): ". . . The moon is virtually aseismic compared with the earth. If it were not for the extremely low seismic noise of the moon, and good sensitivity of the landed seismometers, the moon would appear to be very dead indeed." Compare this with Velikovsky's predictions and the subsequent rationalizations of his proponents.

As to the magnetism in the rocks (which, to Velikovsky's credit, seems to have been a complete surprise to everybody else), his exact words were that the moon's surface "could conceivably be rich in remanent magnetism resulting from strong currents when in the embrace of exogenous magnetic fields." Such a careful wording would have been safe if such magnetism had not been found, but Velikovsky hit the jackpot here. Or did he?

The actual nature of the lunar remanent magnetism is even stranger than Velikovsky had predicted. Far from being a planet-wide field, it is locally "patchy" and varies tremendously in orientation and strength. In fact, the fossil magnetic fields appear related to random impact events, not to the simultaneous imprinting of a uni-directional outside field. The lack of any general homogeneity has led moon geologists to reject the notion of any exterior cause such as the one Velikovsky predicted.

Nor is the timing in agreement. Velikovsky wrote: "I maintain that less than 3,000 years ago the moon's surface was repeatedly molten and its surface bubbled," causing most lunar craters (the others were supposedly created by interplanetary lightning dis-charges which also created the rills). All radioactive dating schemes indicate that the last melting took place billions of years ago, but Velikovsky has found fault with them. Additionally, Velikovsky warned in 1967 of danger to the lives of astronauts: he claimed to have identified a special condition the astronauts most certainly will meet on the moon that may . . . endanger the lives of the astronauts even if they succeed in returning. . . ." This condition was that "radioactivity must still be present on the surface of the moon . . . far exceeding any exposure regarded as safe." Especially near the rayed craters, he forecast "a strong, decidedly harmful, radioactivity." But it wasn't there: Apollo instruments measured no radioactivity more intense than that from terrestrial basalt—and certainly nothing any-where near the danger level.

Astronomy magazine provided a forum for this debate in 1980–81. Responding to my initial anti-Velikovsky sentiments, Leroy Ellen-berger wrote (in part): "Oberg states, for example: 'Radioactive dating (of lunar samples) indicates that the last melting took place billions of years ago. . . .' But radioisotope dating determines age, not thermal history. Regarding the time since the last melting, Velikovsky replied, in response to an earlier critic, 'The question is not when the rocks have been formed or for the first time been crystallized; but when they were heated and partly molten for the last time.' Velikovsky then went on to cite the thermoluminescence study of Apollo-12 cores, which indicated thermal disturbances in historical times."

This is a standard Velikovskian attempt to invalidate *all* radioac-tive dating results. But in citing allegedly supportive "ther-moluminescence data," Ellenberger was walking into a prepared minefield. I zipped my response back off to *Astronomy*: "Radioactive dating *does* determine the age of rocks since last melting. Melting the rocks releases trapped argon atoms which make radioactive decay dating possible."

I also pointed out that the citation of "Apollo-12 thermoluminescence data" (which they claim invalidates these radioactive dating techniques by indicating thermal disturbances within historical times) is simply fallacious. According to Dr. Robert M. Walker, the man who did the original studies, "The discussion of the thermoluminescence data by Velikovsky is completely wrong-headed. . . . Any disturbance we might have reported would have a physical disturbance resulting in the exposure of sub-surface material to solar heating. In fact, the thermoluminescence data prove that Velikovsky's contention—that the moon surface was recently heated—is nonsense." These are the words of the principal investigator of the experiment, whose report was used by Velikovsky as support.

Regarding the actual composition of the lunar surface, Velikovsky was quite specific. In a letter to the National Academy of Sciences on July 2, 1969, he claimed that moon rocks "will be found rich in oxygen, chlorine, sulfur, and iron." Actually, iron was found to be present at or slightly below the average "solar abundance," and both chlorine and sulfur were depleted by several orders of magnitude. Oxygen, while present in great *absolute* amounts, was far from rich in terms of Solar System averages. In fact, since iron was found to exist primarily in the ferric rather than ferrous (or more highly oxidized) forms, it was possible for geologists to demonstrate that lunar rocks were formed in the presence of amounts of oxygen many orders of magnitude *less* than on earth. Velikovsky also predicted that "carbides, into which hydrocarbons would transform when heated, [will be found] in substantial quantities" . . . and they were not. Additionally, he had said that "the moon may well have hydrocarbons in the form of dried naphtha, bituminous rocks, asphalt, or waxes." And again his prediction was wrong.

"In my understanding," Velikovsky wrote in the *New York Times* during the Apollo-11 landing, ". . . less than 10,000 years ago, together with the earth, the moon went through a cosmic cloud of water (the Deluge) and subsequently was covered for several centuries by water. . . ."—yet moonrocks show no trace of such oceans or even a light dew. According to Velikovsky, the moon was captured by the earth, but he departs from other theorists by claiming that this capture occurred in the recent past: "It is probably the most remote remembrance of mankind: the time when there was no moon. . . . The traditions of diverse people offer corroborative testimony to the effect that in a very early age, but still in the memory of mankind, no moon accompanied the earth." That claim, too, has been inconsistent with all relevant space probe data.

The planet Mars played an important role in Velikovsky's reconstruction of ancient history, since it was supposed to have been the agency responsible for a series of worldwide catastrophes in the eighth and seventh centuries B.C. Since Velikovsky expected the planet to have suffered equally or more in the exchange, he assumed it would be covered with fractures—and they would be the explanation for the reported "canals." And so he reported in 1974: "The 'canali' proved to be . . . rifts caused by twisting of strata." But they weren't—the classical canals had been only optical illusions. However, there *were* valleys and chasms on Mars; and however mysterious their origin remains, they do not appear to be the result of "twisting of strata," at least not in the recent past. Missing, too, were the localized areas of strong radioactivity that Velikovsky had prophesied.

"I claimed that neon and argon are chief constituents in the Martian atmosphere," he reminded scientists in 1969, pointing to a passage in *Worlds in Collision* that stated that "neon and argon will be found as main ingredients of the Martian atmosphere." When some highly confused readings from a Soviet Mars probe in 1973 hinted at argon compositions as high as 30 percent, Velikovsky saw himself once again vindicated. However, the celebration was premature, because Viking probes in 1976 determined that argon only made up 1 or 2 percent of the atmosphere—far below Velikovsky's predictions.

Velikovsky's followers urgently sought rationalizations for this discrepancy. Writing in *The Age of Velikovsky*, physicist C. J. Ransom claimed that "it is not now known if this is the average percentage [of argon] . . . or if the Soviet measurements are correct and this is one of the depleted areas," a depletion allegedly caused by the selective concentration of argon in other regions (in other words, the argon was there, but it was *hiding!*). Writing in the pro-Velikovsky journal *Kronos* in 1977, editor-in-chief Lewis M. Greenberg made the excuse that "the relatively low amount *might* be the result of an argon loss to the earth, the moon, and interplanetary space during the celestial events described in *Worlds in Collision* (in other words, the argon had been kidnapped!). Another leading pro-Velikovsky scholar, Ralph E. Juergens (who died a month before Velikovsky did), asserted that the phrase "in rich amounts" must be interpreted in relation to the earth, where argon makes up about 1 percent of the atmosphere, thus leading to the conclusion that Martian argon may be up to twice as plentiful in its relative concentration (in other words, Velikovsky had been right but his prediction had been misinterpreted by everyone including Velikovsky!). Despite all of this waffling, Velikovsky's prediction of "rich amounts" of argon in the Martian atmosphere is still widely trumpeted as yet another vindication of this theories.

A parallel maneuver was used with neon, since Velikovsky had predicted that Mars would have a "neon rich" atmosphere. In describing the Viking experiments, Velikovsky had criticized NASA for anticipating "in my opinion, too little—666 parts per million—neon. . . . The logic that led me to these conclusions was the same that made me make similar advance claims concerning the moon." When Viking landed, it found that the neon concentration was on the order of ten parts per million—almost a hundred times *less* than the figure Velikovsky had forecast was "too little."

Viking had other surprises for Velikovsky. First (and this surprised everyone else, too) was the red sky, despite Velikovsky's explicit prediction in *Worlds in Collision* that "the atmosphere of Mars is invisible . . . [and any Martians] see a black sky, not a blue one as we do." Regarding the possibility of biological activity, he had told an audience in 1974 that ". . . it is not excluded that Mars is richly populated by microorganisms pathogenic to man. . . . I do not discount the probability that the seasonal changes in the color of the Martian surface may be due to seasonal microbial or other low vegetable activity." Although this is very weak for a "prediction," the pattern has emerged that any discovery of life would have been seen as confirmation of another prediction, while the failure to discover life would not *disprove* the "prediction."

The nature of the Martian polar caps also was the subject of some explicit predictions by Velikovsky: "Chances are that they are composed of the same organic molecules as the envelope of Venus," he wrote in 1963, clarifying his earlier prediction that they were carbohydrates ("probably in the nature of carbon," he had said, but *not* just carbon dioxide). Mariner and Viking data took a long time to confirm that the caps contain water ice and frozen carbon dioxide ("dry ice") but there is no evidence for heavier organic compounds.

Jupiter was the source of the birth of Venus and other comets, Velikovsky claimed, and the Red Spot was "an atmospheric effect related to the scar where Venus was ejected." (There is no surface to scar, and the Red Spot is now thought to be a highly stable hurricane that floats freely in the turbulent atmosphere.) "Jupiter must have petroleum," Velikovsky predicted, and "there are some historical indications that Venus—and therefore also Jupiter—is populated by vermin. By "vermin" Velikovsky presumably meant "flies," but all dictionaries extend that term to cover rats, mice, worms, and lice (although not, as Carl Sagan imagined, frogs)—all extremely unlikely denizens of the atmosphere of Jupiter.

Velikovsky's followers are very proud of his prediction of radio noises from Jupiter—which *were* there, much to the surprise of scientists. He made this prediction in a speech given at Princeton on October 14, 1953: "The planet Jupiter is cold, yet its gases are in motion. It appears probable to me that it sends out radio noises as do the sun and the stars." Not only was Velikovsky wrong about the temperature of Jupiter (everyone else was, too), he accepted what turned out to be a mistaken idea that radio point sources in the sky were from stars—they really were from galaxies, as was finally proved about ten years later. The radio noise that Velikovsky predicted, called "thermal" because it is the natural by-product of any object warmer than absolute zero, exists on Jupiter and on other planets as well; but the *characteristic* radio noises from Jupiter are the decimetric (caused by electrons oscillating in the atmosphere) and the decametric (caused by electrical discharges in the upper atmosphere), and these are far more powerful than the thermal radio noises—yet Velikovsky's prediction does not seem to point to them. At best, his "hit" is a glancing one here, and it certainly is not nearly as impressive as later accounts made it sound. (Supporters claim Velikovsky predicted more than just the "thermal" radio noises, and they have some good arguments here.)

Regarding Saturn, Velikovsky claimed that it had once been the largest planet in the Solar System and had constituted a double-planet with Jupiter; Saturn subsequently exploded, showering the Solar System with water, which accounted for the Deluge on earth. "It is conceivable," Velikovsky wrote, "that the earth was, at that time, a satellite of Saturn, afterwards possibly becoming a satellite of Jupiter." The Saturn shower brought earth most of the components of seawater salt: "Chlorine may thus be of extraneous origin," Velikovsky hypothesized. "It could possibly be present in some different combination on Saturn." Also present on Saturn, according to a suggestion made by Velikovsky at NASA's Ames Research Center in August 1972, is "primitive planet life."

Mercury, too, is not ignored. Wrote C. J. Ransom: "Velikovsky believes that Mercury was involved in certain of the earth's recent catastrophic events," based on some unpublished manuscripts he had seen. In public, Velikovsky denounced Einstein's Theory of Relativity as it has been applied to Mercury: "The precession of Mercury is not a relativistic phenomenon," he claimed in 1972, "but results from that planet's electrical charge and its motion in the Sun's magnetic field. . . . Mercury has occupied its orbit only since recent times." Although the Mariner-10 probe made three close fly-by passes of Mercury without detecting any such perturbing force, Velikovsky had earlier suggested that our first moon probes had missed their targets because they had been thrown off course by the moon's electrostatic charge—which conveniently vanished a few years later.

This record of prognostications, if it proves anything, demonstrates that Velikovsky's chief skill had been in making flexible enough predictions, and then in reinterpreting them skillfully in light of ultimate discoveries, in order to produce what superficially does appear to be a remarkable track record. "How could I produce this score of correct prognostications?" he asked in 1974, answering the rhetorical question with the circular assertion that he had been correct from the very beginning.

As we saw earlier, many space scientists have objected to this mode of creating "correct predictions," but Velikovsky would not submit to that type of criticism: "The most despicable of all ways of suppression is denying me the originality and correctness of my predictions," he told a conference at Notre Dame in 1974. Regarding the continuing criticism that his theories had received, he also claimed that "The sons have exerted themselves to outdo their fathers. . . . The scientific community . . . neglected its debt to the public, to truth, and to its own conscience." Echoing that theme, the pro-Velikovskian Italian mathematician de Finetti called the anti-Velikovsky forces a "despotic and irresponsible Mafia." Their criticisms, wrote the pro-Velikovsky journal *Pensee* in 1972, were "chiefly remarkable for dishonesty and incompetence." The level of debate on this issue, obviously, has not been conducive to rationalism and dispassionate consideration of the evidence!

In addition, Velikovsky's predictions did *not* meet the test of scientific validity because they were not formulated so as to be *disprovable*. Instead, phrases such as "could conceivably exist" or "not rule out the probability" were injected into the forecasts, thus allowing nearly any eventuality to be considered either confirmation or null. Some specific predictions were made, and these usually turned out completely wrong—which is why they are not referenced in the pro-Velikovsky literature.

The record of Velikovsky's predictions about space and planetary science therefore provides a sound criterion for judging the validity of both his theories and his method. Both fail.

5

A Personal Reminiscence

Lloyd Motz

In 1950 Dr. Immanuel Velikovsky, an M.D. (psychiatrist) from Moscow University, biblical scholar, Egyptologist, papyrologist, self-taught physicist and astronomer (with many gaps in this phase of his education), linguist, writer, lecturer, and a most stubborn opponent in a debate, published *Worlds in Collision*, a book that stirred a violent tempest in the scientific (particularly, the astronomical) world, and generated a controversy that continued until Velikovsky's death, some thirty years later. This book would probably have caused much less of an uproar, or none at all, if it had been published some ten years earlier or later, if Velikovsky had presented it as primarily speculative rather than as scientific truth, had not affirmed categorically throughout the book that the Newtonian gravitational theory of the dynamics of the Solar System is wrong, and had not insisted that he was the discoverer of a new Solar System dynamics that explained all known Solar System observations, ranging from the surface temperature of Venus to the emission of radio waves from Jupiter.

The scientists' (particularly the astronomers') anger was further compounded by the way the book was published, with particular annoyance at its prepublication publicity, which promised discoveries so startling that the very foundations of science would be shaken, and all of this by a non-scientist, an outsider with whom most laymen, particularly theological fundamentalists and cranks, could easily identify. No wonder the book became an overnight best-seller.

That it was first published by Macmillan, perhaps the most famous publisher at that time of the outstanding authors of physics and astronomy, was considered an affront by many prestigious scientists, and a violent campaign, led by the late famous Harvard astronomer, Harlow Shapley, was directed against Macmillan, urging it to cease publication of *Worlds in Collision,* to which Macmillan agreed, under threat of being cut off from any new book manuscripts from the science community. Fearful of irreparable financial damage to Macmillan and himself, Velikovsky accepted Doubleday's offer to take the book off Macmillan's hands and reissue it under its own imprint. The only people who suffered under this arrangement were Gordon Atwater, then director of the Hayden Planetarium and Chairman of the Astronomy Department of the American Museum of Natural History, and James Putnam, associate editor of Macmillan; they were both summarily dismissed from their positions. Atwater had advised Macmillan to publish the book and had strongly touted and defended it in print before and after its appearance, and Putnam had accepted it for publication very enthusiastically and thus became Macmillan's natural whipping boy. Some of the onus for the book's publication fell as well on Eric Larrabee, author; John Lear, author; John T. O'Neill, science editor of the now defunct *New York Herald Tribune;* and Dr. Horace M. Kallen, a famous philosopher—people not directly involved in the book's publication but who had praised it warmly. Already in 1946, O'Neill referred to Velikovsky's manuscript before its publication as "a magnificent piece of scholarly historical research;" Larrabee and Lear wrote articles in *Harper's* and *Collier's,* announcing Velikovsky's work as a major discovery, ranking with the work of Darwin, Newton, and Einstein; and Kallen praised its originality.

That the book appeared, with its challenge to scientific authority, shortly after the dropping of nuclear bombs on Hiroshima and Nagasaki must have greatly influenced the scientific community in its bitter and violent denunciation of Velikovsky—a denunciation labeling him a charlatan with no regard for truth, seeking only self-aggrandizement. The scientists who condemned Velikovsky's astronomy as trash sought the aid of historians and biblical scholars to strip Velikovsky's historical research of any merit, although it was clear to everyone who read Velikovsky's writing that his historical research was very sound; but his astronomy and science in general were so faulty and imprecise that his worthy researches were not given and have still not been given the attention they deserve.

As I indicated, the tenor of the times in 1950 was hardly conducive to quiet acceptance of what many scientists considered a threat to their victory in a long struggle against ignorance and irrationalism. The production of the nuclear bomb and the direct evidence of its power had elevated scientists to unheard-of social, political, and military levels. Many of their names were household words, and they became unfathomable but infallible intellectual figures to whom no problem was insoluble. But already, second thoughts about the morality of the bombing of Hiroshima and Nagasaki were casting doubts about the human qualities of scientists, and now Velikovsky was questioning the validity of their science, not at some minor level, but on the very highest plane—the law of Newtonian gravity as applied to planetary motions. If accepted by the public, this could lead to a backlash that would threaten scientific progress at every level. Hence the need to denounce Velikovsky's work and all he stood for if science and scientists were not to suffer a severe decline in the public's estimate. Velikovsky was to state some years later, in his own evaluation of the cause of the scientists' intense opposition to his theories, that ". . . I was . . . carrying my heresy into the most sacred field, the holy of holies of science, to celestial mechanics. . . ."

Velikovsky's popularity and the great appeal of his books held for the public stemmed not only from his challenge to science but also from his scholarly credentials, which placed him far above the common, run-of-the-mill cranks who continuously bombard the public with tracts supporting such nonsensical ideas as a flat or hollow earth, astrology, perpetual motion machines, anti-gravity devices, and the like. Velikovsky's credentials were not those of a scientist, but they were formidable and overwhelming to a non-scientist. He was described, in the publicity for *Worlds in Collision,* as "an editor, historian, and physician with an incredible range of competence in the sciences." Born in 1895 in Russia, he had, indeed, studied, though not in equal depth, a wide range of subjects, including the natural sciences, economics, law, history, medicine, and pychoanalysis at universities in Edinburgh, Kharkov, Moscow, Vienna, and Zurich; he was thus, quite properly, considered to be a "universal student." This very scholarliness, of course, contributed to the anger and antagonism of the opposing scientists, for they felt that Velikovsky was using his scholarly stature to perpetrate a scientific fraud and thus to elevate himself to the ranks of the greatest minds. But in this evaluation of Velikovsky's motives, attitude, and philosophy, they were wrong.

Velikovsky had not set out to challenge Newtonian authority and to invent a new cosmology, or even to question some features of Newtonian gravity as applied to the motions of the planets. Indeed, with a meager background in mathematics and astronomy and hardly any in physics, he was ill prepared for such a project, to say the least. Already established in Europe as an author and a noteworthy scholar when he came to New York in 1939, he planned only to carry on his historical researches in the Columbia University libraries. His interests turned to ancient history, particularly as developed in Egyptian papyri, biblical sources, and the mythologies and folklores of various civilizations. According to his own account, he spent ten years comparing these historical sources, and concluded that there is a chronological discrepancy between the biblical portrayal of the exodus and the portrayal in the Ipuwer Papyrus of what Velikovsky accepted as the same series of events, a series of catastrophes that struck the Egyptians. To validate this conclusion,

Velikovsky had to bring Egyptian history into coincidence with biblical history and that required advancing Egyptian chronology by
about five hundred years. This was the starting point of *Worlds in
Collision* and his cosmological adventures, which he embarked upon
for historical rather than scientific reasons. He describes his reasons
for his "great adventure" and state of mind at that time in a very
revealing passage in the preface of *Worlds in Collision:*

> It was in the spring of 1940 that I came upon the idea that in the
> days of the Exodus, as evident from many passages of the Scriptures,
> there occurred a great physical catastrophe, and that such an event
> could serve in determining the time of the Exodus in Egyptian history or
> in establishing a synchronical scale for the histories of the peoples
> concerned. Thus I started *Ages in Chaos*, a reconstruction of the history
> of the ancient world from the middle of the second millennium before
> the present era to the advent of Alexander the Great. Already in the fall
> of that same year, 1940, I felt that I had acquired an understanding of the
> real nature and extent of that catastrophe, and for nine years I worked
> on both projects, the political and the natural histories. Although *Ages in
> Chaos* was finished first, in the order of publication it will follow this
> work.
>
> *Worlds in Collision* comprises only the last two acts of the cosmic
> drama. A few earlier acts—one of them known as the Deluge—will be the
> subject of another volume of natural history.

It is clear from this passage that his principal concern initially was to
set the historical record straight as far as the Exodus from Egypt is
concerned. But he was soon led to geological and astronomical conclusions that are completely at variance with the the gravitational
dynamics of the solar system, and he thus decided to construct his
own planetary dynamics. A hint of this is given in the third sentence
of the above passage where he states that in the fall of 1940 he "had
acquired an understanding of the real nature and extent of that
catastrophe . . ."

Here his interpretation and understanding of "that catastrophe"
went far beyond the Exodus—to him it meant something global, which could have its explanation only in a series of vast celestial
phenomena. He was drawn to that conclusion by his "discovery"
that all civilizations at the time of the Exodus described catastrophic events, and these total global upheavals greatly influenced his
thinking, as indicated by the following paragraphs from *Worlds in
Collision:*

> The historical-cosmological story of this book is based on the evi-
> dence of historical texts of many peoples around the globe, on classical
> literature, on epics of northern races, on sacred books of the peoples of
> the Orient and Occident, on traditions and folklore of primitive peoples,
> on old astronomical inscriptions and charts, on archaeological finds,
> and also on geological and paleontological material.

If cosmic upheavals occurred in the historical past, why does not the human race remember them, and why was it necessary to carry on research to find out about them? I discuss this problem in the Section "The Collective Amnesia." The task I had to accomplish was not unlike that faced by a psychoanalyst who, out of disassociated memories and dreams, reconstructs a forgotten traumatic experience in the early life of an individual. In an analytical experiment on mankind, historical inscriptions and legendary motifs often play the same role as recollections (infantile memories) and dreams in the analysis of a personality.

Can we, out of this polymorphous material, establish actual facts? We shall check one people against another, one inscription against another, epics against charts, geology against legends, until we are able to extract the historical facts.

In a few cases it is impossible to say with certainty whether a record or a tradition refers to one or another catastrophe that took place through the ages; it is also probable that in some traditions various elements from different ages are fused together. In the final analysis, however, it is not so essential to segregate definitively the records of single world catastrophes. More important, it seems, is to establish (1) that there were physical upheavals of a global character in historical times; (2) that these catastrophes were caused by extraterrestrial agents; and (3) that these agents can be identified.

The last three lines in the final paragraph of this quotation are the key to Velikovsky's transition from a scholar and historian to a purveyor of a new theory of planetary motions; he had to find some kind of celestial phenomenon that had an important (that is, observable on a grand scale) impact on the geological history of the earth and whose epoch and span of time coincided with the global catastrophes described in the bible (e.g., the Exodus, Joshua's command that "the sun stand still," etc.) and in folklore and tales from all over the globe (Mayan, Egyptian, Greek, Finnish) of similar events, such as fires, earthquakes, and floods. He buttressed his arguments for the need of a non-Newtonian Solar System dynamics by pointing to the many questions about the sun and planets that are still unanswered. In particular, he emphasized the present apparent conflict between modern theories of gradual evolution and catastrophism and placed himself squarely on the side of catastrophism and, in particular, in favor of "cosmic collisions" as "implicit in the dynamics of the universe."

Having arrived at this point of view Velikovsky sought unassailable "historical evidence" for the causes of the coincident global catastrophes as described in prehistoric tales and legends, and found what he accepted as such evidence in the ancient records of the apparent motions of Venus among the constellations and its "strange appearance," as recorded by observers in various ancient civilizations. He discovered, according to his own account, that Venus is not mentioned in the Egyptian papyri that were written before the Exodus nor in the literature, in that period, of other civilizations. Venus' first recorded appearance came, again according to Velikovsky, only with the onset of the catastrophes. Moreover this literature, taken from hundreds of sources by Velikovsky, describes Venus not as one of five known planets but as a strange extended object. From these descriptions and the datings he concluded that Venus is a new planet, born about 1500 B.C., first as a comet that, in that form, produced all the recorded catastrophes in a series of encounters with the earth. Only some hundred years later, in this Velikovskian history, did Venus shed its comet trappings to become a well-behaved planet. From this "birth of Venus," with incredible ingenuity and disregard of basic scientific principles, did Velikovsky develop a new version of geologic catastrophism and a new dynamics of the solar system in which gravity plays a minor role. These hypotheses also led him to novel suggestions on an almost incredible range of subjects, from the identity of the Queen of Sheba to the cause of ice ages; from a new theory of evolution to the origin and nature of the manna that fed the wandering Israelites; from the ejection of Venus from Jupiter to radio waves from Jupiter; and so on. To build his case Velikovsky quotes from innumerable sources culled from many ages and many parts of the world; thus *Worlds in Collision* contains references to some five thousand sources and over three hundred correlations and deductions.

With little understanding of the severe restrictions that the basic conservation principles (the conservation of energy, momentum, and angular momentum) impose upon the motions of bodies in a dynamical system such as the solar system, and a profound belief that he had unraveled a great historical mystery and thus had found the answers to some very perplexing geologic, geographic, and astronomical questions, it was easy for Velikovsky to jump from the apparent gaps and discrepancies in the ancient data about Venus to the conclusion and unshakable belief that Venus is a young planet, ejected some three thousand five hundred years ago as a huge comet by a great cataclysm on Jupiter. This idea struck him like a thunderbolt and impelled him to search for supporting evidence for his bizarre hypothesis wherever he could. If he was to challenge scientific facts and theories that had stood the test of hundreds of years of careful observations and had great predictive powers, he had to play the game of the scientists, which meant far more than just making some broad statements about biblical stories, Egyptian papyri, collective amnesia, and calendar discrepancies. To this end Velikovsky pursued three distinct courses: (1) develop a new theory of geologic catastrophism as against gradualism; (2) present a series of astronomical and geological deductions that stem from his hypothesis; (3) learn some basic science.

Fully aware of the great gap in his knowledge and understanding of physics and astronomy, Velikovsky came to me in the early part of 1951, before *Worlds in Collision* appeared, to ask that I tutor him in astronomy, a request that I had to reject. I agreed, however, to discuss his ideas critically and point out where he was wrong. This led to a series of interviews, telephone conversations, exchanges of letters, debates, and confrontations in journals. Two things were immediately clear to me: (1) Dr. Velikovsky was neither charlatan nor crank, but a dedicated, brilliant scholar whose historical investigations led him to his non-Newtonian Solar System of dynamics, to which he stubbornly adhered until his death; (2) he had only the vaguest understanding of such basic physical principles as conservation of angular momentum, gravity, and entropy.

When I pointed out to him that his description of Venus as an errant, unpredictable, meandering comet conflicts with and indeed contradicts the principles of conservation of energy and of angular momentum, he argued that historical evidence must guide us in our acceptance or rejection of scientific principles, and if there is a conflict between a scientific theory and history, so much the worse for the theory. If, he insisted, Newton's laws cannot explain the historical evidence about Venus' behavior, then Newtonian gravity, as the governing force in the solar system, must be replaced by another force, and he believed that he had found what he wanted in the electromagnetic force that, he insisted, could account for the "historical astronomical facts." He made the point that Newton had developed his gravitational theory of the dynamics of the Solar System long before the electromagnetic force was understood or even known in more than a very superficial way; hence, he argued, gravity plays a minor role in the Solar System and, taken alone, leads to an incomplete description of planetary motions.

With this bold but incorrect idea to "guide" (really misguide) him and in spite of our many discussions and his gradual acceptance of basic physical principles, he persisted in maintaining his overall thesis of general catastrophism rather than gradual evolution as the prime cause of changes on the earth and in the heavens. Since, according to Velikovsky, the sudden birth of Venus caused this catastrophism, this birth had to be dramatic enough to call the attention of all the peoples of the world to it and had to be accompanied by enough energy to do all the remarkable things Velikovsky ascribed to it. And here mythology came to his aid with its many references to powerful Jupiter and his "thunderbolts" and to the birth of Athene from "the head of Jupiter" in a "ball of fire." Accepting this fanciful picture as a literal description of actual celestial events, Velikovsky based his entire thesis on the idea, which he accepted as a fact, that "Venus was expelled (from Jupiter) as a comet and then changed to a planet after contact with a number of members of the solar system." This led him to predictions about Venus, the moon, the earth, Mars, and Jupiter some of which were later verified. Thus, he reasoned, that since Venus was born in a burst of flame only a few thousand years ago, its surface still had to be hot since it had no time to cool off, and Jupiter, still in a state of violent activity, owing to its "painful" ejection of Venus, must be a strong source of radio waves. The subsequent observational discovery by K. Franklin that Jupiter emits intense radio waves, and the further discovery, from the intensity of radio emissions from the surface of Venus, that the temperature of that surface must be about 800° F, convinced Velikovsky that his theory of the origin of Venus and his reconstruction of its history were correct. It was, of course, obvious to scientists that this conclusion was completely unjustified, for the successful prediction of an event does not validate the thesis on which that prediction is based. This was, indeed, the essential point at issue between Velikovsky and his critics. That correct deductions can be drawn from false premises is too obvious a truth to be argued, but this simple point either escaped Velikovsky or he refused to accept it. In any case, he built a vast pyramid of predictions on it and gathered every bit of information he could find in astronomical and geological literature that lent credence to that pyramid.

Since the scientific community on the whole was convinced that Velikovsky's researches had no merit at all, whether historical or scientific, it completely disregarded his predictions, for which he was given no credit and which were never referred to or mentioned in the literature. It is no wonder, then, that Velikovsky became deeply embittered, frustrated, and accusatory. Whether his basic assumptions were correct or not, he felt that the truth of his predictions should be recognized and his prior rights to them acknowledged. As a last resort in his battle for recognition he appealed for fair treatment directly to individual scientists, among them the well known Princeton physicist, V. Bargmann and myself. Recognizing the justice of his priority claims we therefore wrote a letter to the AAAS journal *Science* whose last paragraph stated our position as follows:

> Although we disagree with Velikovsky's theories, we feel impelled to make this statement to establish Velikovsky's priority of prediction.
>
> . . . a careless reading of Eric Larrabee's article may leave the unwary reader with the false impression that Dr. Bargmann and I accept and agree with Dr. Velikovsky's ideas. . . . I do not support Velikovsky's theory but I do support his right to present his ideas and to have these considered by responsible scholars and scientists as the creation of a serious and dedicated investigator and not the concoctions of a charlatan seeking notoriety.
>
> . . . Dr. Velikovsky's ideas do not constitute a new theory since they contain no new fundamental principles of nature. . . .
>
> That there is no astronomical evidence for electromagnetic forces of the magnitude required by Velikovsky's theory . . . and that such forces of the required magnitude . . . would destroy the . . . completely verified laws of planetary motion are not accepted by Dr. Valikovsky as valid arguments against his ideas. Since . . . these . . . have led him to certain predictions . . . he is convinced that his ideas must be right. But . . . verified predictions alone do not validate a theory, and my position is that nothing has happened during the last decade to make Velikovsky's theory any more acceptable now than . . . when . . . first published. . . .

. . . however, . . . his predictions should be recognized and . . . his writings . . . carefully studied and analyzed because they are the product of an extraordinary and brilliant mind, and are based on some of the most concentrated and penetrating scholarship and research of our period. . . . Dr. Velikovsky has performed a service to science in collecting the vast amount of data . . . and bringing clearly to the attention of the scientific community the many discrepancies that exist in our understanding of the history of our earth during the last geologic period."

This letter, appearing as it did in the December 12, 1962, issue of *Science*, when the intense heat of the Velikovsky battle had been dissipated, was used as ammunition by the pro-Velikovsky forces to continue their hopeless battle to overthrow Newton. Since our position and point of view were misinterpreted or deliberately distorted by Velikovsky's supporters, I wrote a letter to *Harper's* which stated our position clearly; while pointing out the untenability of Velikovsky's physics, I still appealed for a fair evaluation of Velikovsky's scholarly contributions.

Now that the Velikovsky affair is rapidly being forgotten and losing its ability to excite antagonisms, we can evaluate dispassionately its impact on science and on our thinking in general. As far as science is concerned, its influence was nil, for Velikovsky's entire hypothesis rests on his false assumption that Venus was ejected from Jupiter as a comet. As I have already shown in a rebuttal to Velikovsky that appeared in the *Yale Journal* following a Princeton debate between Velikovsky and myself, Velikovsky's Venus-Jupiter hypothesis is wrong—in fact, impossible from the very start. For Jupiter to have ejected from its interior a mass equal to Venus' at a speed needed to launch it into its present orbit around the sun, Jupiter would have had to release or expend in a matter of seconds or minutes as much energy as our sun emits in more than a year. Jupiter would therefore suddenly have appeared as bright as a million suns emitting enough energy to vaporize all the planets near it, especially the earth. This alone shows the absurdity of Velikovsky's claim; but there are other arguments against it as devastating as this which I shall not go into. In any case, all of Velikovsky's "remarkable predictions" have been fully explained by the standard physical theories without calling upon Venus.

What about Velikovsky's historical research that triggered his attack on the Newtonian gravitational dynamics of the Solar System? Was it honest research or deliberately falsified to support his wrong astronomical assumptions and his electromagnetic Solar System? The scientists who had organized the violent and bitter attack against Velikovsky's cosmology also did all they could to discredit his historical research, calling upon such prominent historians as O. Neugebauer at Princeton to show that Velikovsky's ancient chronology is incorrect. The late noted astronomer Cecilia Payne-Gaposchkin, who was somewhat of an authority in the history of science, accused Velikovsky directly of misquoting the Egyptian papyri and other ancient historians such as Herodetus; Neugebauer attacked him on similar grounds. Later evidence showed that Payne-Gaposchkin had misrepresented Velikovsky's historical theories to make them appear ludicrous. Similarly, Neugebauer's criticism appeared to be without merit.

At the present time some of the evidence about the chronology of ancient events seems to favor Velikovsky. In particular, some historians believe he was right on the age of the Meso-American civilization and the dating of Tutankhamen, and thus performed an important service in calling attentiion to these discrepancies in ancient chronology. But his being right in these instances does not validate his astronomy.

In a recent book *The Cosmic Serpent* the British astronomers Victor Clube and Bill Napier refer to Velikovsky's work as a "remarkable piece of historical analysis" and that his identification of the Queen of Sheba with the Egyptian Queen Hatshepsut, who visited King Solomon in the "legendary land of Punt" (now Palestine), is a "significant and remarkable achievement" and that it has not been given the attention by experts that it deserves." They go on further to say that "there was a widespread anticipation of an encounter of the earth with a comet or its debris in 687 B.C. and this event could have been as he (Velikovsky) suggests, a significant turning point in the history of civilization."

One final point in favor of Velikovsky is his emphasis on the important role of catastrophe in geologic processes, which is now generally accepted. Here, too, his historical analysis was decisive in bringing him to that position.

In the final analysis, the Velikovsky affair did not diminish but enriched the history of science, revealing, as it did, the frailty of the tolerance of the scientific community. Velikovsky's only "sin" was in trying to promote a faulty Solar System dynamics, but his speculations are meek and really well within the bounds of accepted scientific theorizing compared to what is going on in cosmology and particle physics today. Honest errors and speculation in the pursuit of science are absolutely essential; without them there can be no progress. We physical scientists should therefore welcome, rather than revile, people like Velikovsky who, periodically, force us to reexamine our basic assumptions.

6

Velikovsky's Historical Revisions

WILLIAM H. STIEBING, JR.

Professor Motz's account of the Velikovsky Affair emphasizes a point that is not often recognized by Velikovsky's scientific critics, namely that Velikovsky's thesis was essentially *historical*, not scientific. Velikovsky became convinced that the Exodus story in the Bible recorded an ancient catastrophe of such magnitude that it should have been noted in the writings, folklore, myths, and legends of many other peoples. After much research into the literature and mythology of various ancient nations, Velikovsky located a number of supposed references to catastrophes, and from them he created his scenario of colliding planets.[1] If scientific theory could not be reconciled with the historical evidence, he asserted, then the scientific theory is wrong.

One Egyptian text in particular, the Ipuwer Papyrus, seemed to Velikovsky to be a contemporary Egyptian account of the cataclysmic events associated with the Exodus. But since Egyptologists dated the Ipuwer text much earlier than the Hebrew Exodus, Velikovsky decided that Egyptian chronology had to be advanced by some five hundred to six hundred years.[2] However, despite the recent favorable comments by astronomers Victor Clube and Bill Napier,[3] Velikovsky's revised chronology and historical reconstructions were as erroneous as were his ideas of celestial mechanics.

Velikovsky's methodology for locating ancient references to cosmic catastrophes was very subjective and highly questionable. A historian's first duty is to examine his sources to determine their date and reliability. Velikovsky never did this. He assumed that Medieval Jewish and Arab legends dealing with the Exodus had the same historical value as the biblical account written many hundreds of years earlier. And he uncritically accepted the biblical account itself as an accurate description of events at the time of the Exodus, even though biblical scholars have contended that it is a composite narrative, the earliest elements written some two hundred years after the time of Moses and the latest portions dating from the time of the Babylonian Exile, about six hundred years after the Exodus. In other instances Velikovsky identified characters from myths with the planets Venus or Mars whenever the account would seem to fit his scenario of clashing planets, but usually provided no evidence that the creator of the myths intended such identifications. And his interpretation of many ancient texts was faulty because of his incomplete understanding of the languages and cultures involved. We cannot detail the problems with Velikovsky's methodology here because of space limitations,[4] but a couple of examples will suffice to illustrate the point.

He asserts that Mexican Aztec stories about the ferocity of their war god derived from catastrophes of the eighth and seventh centuries B.C. when Mars supposedly had several encounters with Venus and earth. He also claims that these cosmic events caused the Aztecs to change their homeland and that the appearance of heavenly warfare between Mars and Venus led the Aztecs and Toltecs to engage in combat, each tribe fighting on behalf of its patron deity.[5] However, there were no wars between the Toltecs and Aztecs. The Toltec empire ended in the twelfth century A.D. and the Aztec empire was not created until the fifteenth century A.D. Moreover, there is no evidence that the Aztecs even existed as an organized tribe in the seventh century B.C. The earliest event that can be connected with the Aztecs is their legendary migration from their place of origin (Aztatlan) to Mexico City (Tenochtitlan), which seems to have occurred between A.D. 1111 and 1345.[6] How can legends describing this movement possibly provide information on a cosmic cataclysm that supposedly took place in 687 B.C., more than seventeen hundred years earlier?

Velikovsky tended to interpret poetry and myths as literal statements of fact. Poetic exaggeration and metaphorical or symbolical meanings are not considered as explanations for the passages he quotes. If one took the same literalistic approach to modern poetry as exemplified in popular songs of the last few decades, one could find references to catastrophes on every hand. Do you recall experiencing the worldwide cataclysms referred to in "Blue Moon," which describes the moon changing color from blue to gold (obviously a passing "comet" heated it to incandescence); or "The First Time Ever I Saw Your Face," which states that the earth moved like the trembling heart of a captive bird (indicating earthquakes all over the earth's surface); or "On the Street Where You Live," which claims that the poet suddenly was swept into the air to a height of several stories (due to the hurricane-force winds generated by the approach of some cosmic body)?

Another problem with Velikovsky's interpretations of ancient texts and myths is that much of the evidence that he cited to support his belief that cosmic catastrophes occurred during the fifteenth and seventh centuries B.C. could be so used only if his chronological revisions were correct. And they were not.

Radiocarbon (C-14) dating would seem to offer an objective test of Velikovsky's revised Egyptian chronology. At first Velikovsky pleaded with scientists to perform C-14 tests on Egyptian Eighteenth Dynasty material, thinking such tests would prove that his chronology was correct.[7] Later he challenged the validity of this dating method since it was based on the assumption that the amount of carbon-14 in the atmosphere has remained relatively constant throughout the ages. However, the cosmic cataclysms that Velikovsky postulated for the fifteenth and seventh centuries B.C. would have changed the ratio of normal carbon and radioactive carbon for those periods.[8] So if Velikovsky's theory were correct, radiocarbon dates for material from the seventh century B.C. and earlier would be erroneous.

In recent years radiocarbon dates have been corrected by testing tree rings from bristlecone pines in the southwestern United States and Irish oaks in Ireland. These tests have shown that the amount of carbon-14 in the atmosphere did not suddenly increase or decrease in the fifteenth and seventh centuries B.C. as it should have according to Velikovsky's scenario. Furthermore, when the radiocarbon dates for Egyptian Eighteenth Dynasty material are calibrated using the results from the tree ring tests, they generally support the conventional chronology, not that of Velikovsky.[9]

Archaeological stratigraphy also indicates that Velikovsky's chronology is wrong. The archaeological and historical evidence against Velikovsky's dating scheme has been given in more detail elsewhere.[10] Here, let me summarize just one point. Velikovsky did not attempt to revise the chronology of the Israelite monarchy, which is tied to the detailed chronology of Mesopotamia. In Palestinian archaeological terminology, the period of the Hebrew monarchies, the Assyrian Empire, and the subsequent Neo-Babylonian Empire is the Iron Age. According to the revised chronology, the Iron Age is also the era of the Egyptian Eighteenth and Nineteenth Dynasties. But stratified archaeological deposits in Palestine-Syria indicate that the Egyptian Eighteenth and Nineteenth Dynasties were contemporary with the Palestinian Late Bronze Age, not the Iron Age! Late Bronze Age occupation layers containing material from the Eighteenth and Nineteenth Dynasties of Egypt are found *below* Iron Age strata. The stratified physical evidence at archaeological sites will not allow one to move Egyptian chronology forward by more than five hundred years while keeping Israelite and Mesopotamian chronology at rest!

Prehistoric Greek chronology is also linked to Egypt by archaeological deposits. Large numbers of Mycenaean pottery vessels were uncovered at Tell el-Amarna, capital of the Egyptian Eighteenth Dynasty pharaoh Akhenaton. Velikovsky recognized this and realized that when he advanced Egyptian chronology by some five hundred years he was also moving up the Mycenaean Era by the same number of years. This would eliminate the "Dark Age" that was conventionally thought to lie between the Greek Mycenaean Age and the Archaic Era.[11] However, Mycenaean pottery is also frequently found in archaeological deposits in Syria and Palestine. But it does *not* occur in Iron Age II layers where it should according to the revised chronology. Instead, it is found in Late Bronze strata. Greek protogeometric and geometric pottery occurs in Palestinian-Syrian Iron Age strata. And these pottery styles are found in layers *above* Mycenaean remains at Greek sites. It may seem easy to eliminate the Greek Dark Age and move directly from the Mycenaean Period to the Archaic Era when formulating a chronology, but the physical evidence of archaeology cannot be disposed of so easily. The Geometric Period (the "Dark Age") follows the Mycenaean Age and precedes the Archaic Period in Greece, and the Geometric Period corresponds to the Palestinian Iron Age. The earlier Mycenaean Age and the Egyptian Eigthteenth and Nineteenth Dynasties were contemporaneous with the Palestinian Late Bronze Age.

Velikovsky does not seem to have ever fully understood this argument, and he never really answered it.[12] But some of his supporters have recognized its validity. Hatshepsut cannot be made a contemporary of Solomon or Akhenaton moved to 850 B.C. unless one also revises the correlations between stratigraphic archaeological deposits and biblical events. So some supporters of Velikovsky's chronology have attempted to revise the accepted dates for Palestinian archaeological periods.[13] But their schemes have also failed to stand the test. By correlating the Late Bronze Age with the Israelite monarchy, they introduce problems between the archaeological evidence and biblical accounts of an era for which the Bible has proven to be quite accurate.[14] Egyptian chronology cannot be revised as radically as Velikovsky suggested. And without such a chronological revision, most of the textual and archaeological evidence for cosmic catastrophes in the fifteenth and seventh centuries B.C. is invalid.

NOTES

1. I. Velikovsky, *Worlds in Collision* (New York: Macmillan, 1950), viii.
2. I. Velikovsky, *Ages in Chaos* (Garden City, New York: Doubleday, 1952), 12–53.
3. V. Clube and B. Napier, *The Cosmic Serpent* (New York: Universe Books, 1982), 226–237.
4. For more detail on these and other points concerning I. Velikovsky's ideas see W. Stiebing, Jr., *Ancient Astronauts, Cosmic Collisions, and Other Popular Theories About Man's Past* (Buffalo, New York: Prometheus Books, 1984), 57–80.
5. I. Velikovsky, *Ages in Chaos*, 253–254, 269.
6. N. Davies, *The Aztecs: A History* (Norman, Oklahoma: University of Oklahoma Press, 1973), 8; and N. Davies, *Voyagers to the New World* (New York: William Morrow, 1979), 177.

7. See the letters of I. Velikovsky published in "ASH," *Pensee* 4, no. 1 (Winter 1973–74): 5–19.

8. I. Velikovsky, "The Pitfalls of Radiocarbon Dating," *Pensee* 3, no. 2 (Spring–Summer 1973): 12–14, 50.

9. C. Renfrew, "Carbon-14 and the Prehistory of Europe," *Scientific American* 225, no. 4 (Oct. 1971): 66–67; C. Renfrew, *Before Civilization: The Radiocarbon Revolution and Prehistoric Europe* (Cambridge: Cambridge University Press, 1979), 69–82; and I. Shaw, "Egyptian Chronology and the Irish Oak Calibration," *Journal of Near Eastern Studies* 44, no. 4 (Oct. 1985) 295–317.

10. W. Stiebing, Jr., "A Criticism of the Revised Chronology," *Pensee* 3, no. 3 (Fall 1973): 10–12; and *Ancient Astronauts, Cosmic Collisions*, 75–76.

11. I. Velikovsky, *Ages in Chaos*, 180–182.

12. See I. Velikovsky, "A Reply to Stiebing," *Pensee*, 4, no. 5, (Winter 1974–75): 24, 49. The arguments presented there miss the point as is noted by W. Stiebing, Jr., "Rejoinder to Velikovsky," *Pensee*, 4, no. 5, (Winter 1974–75): 24–26 and J. Bimson, "Can There Be a Revised Chronology Without a Revised Stratigraphy?," *Ages in Chaos?* (Proceedings of the Residential Weekend Conference, Glasgow, 7–9 April 1978), *Society for Interdisciplinary Studies Review*, 6, nos. 1–3 (1982): 16.

13. Bimson, "Can There Be a Revised Chronology?" 16–26 and D. Courville, *The Exodus Problem and its Ramifications*. (Loma Linda, Ca.: Challenge Books, 1971).

14. W. Stiebing, Jr., *Ancient Astronauts, Cosmic Collisions*, 75–80 and "Should the Exodus and the Israelite Settlement be Redated?," *Biblical Archaeology Review*, 11, no. 4 (July/Aug. 1985): 58–69.

REFERENCES

"ASH." *Pensee* 4, no. 1, Winter 1973–74: 5–19.

Bimson, J. "Can There Be a Revised Chronology Without a Revised Stratigraphy?" *Ages in Chaos?* (Proceedings of the Residential Weekend Conference, Glasgow, 7–9 April 1978), *Society for Interdisciplinary Studies Review*, 6, nos. 1–3, (1982): 16–26.

Clube, V. and B. Napier. *The Cosmic Serpent*. New York: Universe Books, 1982.

Courville, D. *The Exodus Problem and Its Ramifications*. Loma Linda, California: Challenge Books, 1971.

Davies, N. *The Aztecs: A History*. Norman, Oklahoma: University of Oklahoma Press, 1973.

Renfrew, C. "Carbon-14 and the Prehistory of Europe." *Scientific American* 225, no. 4 (Oct. 1971): 63–72.

Refrew, C. *Before Civilization: The Radiocarbon Revolution and Prehistoric Europe*. Cambridge: Cambridge University Press, 1979.

Shaw, I. "Egyptian Chronology and the Irish Oak Calibration." *Journal of Near Eastern Studies* 44, no. 4 (Oct. 1985): 295–317.

Stiebing, Jr., W. "A Criticism of the Revised Chronology." *Pensee* 3, no. 3 (Fall 1973): 10–12.

Stiebing, Jr., W. "Rejoinder to Velikovsky." *Pensee* 4, no. 5 (Winter 1974–75): 24–26.

Stiebing, Jr., W. *Ancient Astronauts, Cosmic Collisions, and Other Popular Thories About Man's Past*. Buffalo, New York: Prometheus Books, 1984.

Stiebing, Jr., W. "Should the Exodus and the Israelite Settlement be Redated?" *Biblical Archaeology Review* 11, no. 4 (July/August 1985): 58–69.

Velikovsky, I. *Worlds in Collision*. New York: Macmillan, 1950.

Velikovsky, I. *Ages in Chaos*. Garden City, New York: Doubleday, 1952.

Velikovsky, I. "The Pitfalls of Radiocarbon Dating." *Pensee* 3, no. 2 (Spring–Summer 1973): 12–14, 50.

Velikovsky, I. "A Reply to Stiebing." *Pensee* 4, no. 1, (Winter 1973–74): 38–42.

Velikovsky, I. "A Concluding Retort." *Pensee* 4, no. 5 (Winter 1974–75): 24, 49.

7

The Velikovsky Affair: Science and Scientism in Contemporary Society

HENRY H. BAUER

Introduction

The Velikovsky affair has become something of a classic in the history of science. For social scientists it functions as an exemplar of the resistance offered by establishment science to revolutionary ideas proposed by outsiders. For scientists, particularly astronomers and physicists, it is a frightening example of how gullible the media and the public can be toward pseudo-scientific notions. The affair is worth recalling also as an object lesson for scientists in how not to participate in public debates; and it illustrates the mistaken notions that are rampant in all sections of our soceity about the true nature of the scientific enterprise.

All these aspects of the affair have to do with our contemporary approach to acquiring knowledge, and reveal that scientistic belief is virtually universal: that is, the belief that science is the ultimate arbiter of truth. However, because science does not offer satisfying answers to questions of human significance, its authoritative role is resented, and such challenges to establishment science as offered by Velikovsky readily find a hearing.

Needed is a more realistic appreciation of the nature of the scientific enterprise, in particular of the limits of the domain to which it is applicable. Such an understanding would then leave room for the religion for which we pine and for which science has not proven a satisfactory substitute.

Brief History of the Affair

The Velikovsky affair exploded in 1950 with the publication of *Worlds in Collision*. The book drew inferences from history, legend, and myth to conclude that a comet had erupted out of Jupiter and had come into close approach with the earth about 1500 B.C., producing cataclysmic effects described in the biblical story of the Exodus and in other ancient sources. There were further catastrophic encounters between the comet, earth, and Mars in the eight century B.C., before the comet settled into orbit around the sun as what is now the planet Venus.

The popular media acclaimed the originality and scholarly virtues of Velikovsky's book; but the academic community, particularly the astronomers, expressed outrage that such pseudo-scientific nonsense could be given any credence in this modern, supposedly scientific age. The publisher, Macmillan, was put under boycott, and the book's editor lost his job there. Gordon Atwater was dismissed from the Hayden Planetarium of the American Museum of Natural History for suggesting that Velikovsky be taken seriously. But the book continued to sell very well under the Doubleday imprint, and public controversy continued for many months before abating.

More than ten years later the affair erupted again, because Velikovsky claimed to have been vindicated: astronomers had now discovered what he had predicted—Venus was much hotter than anticipated by scientists, the earth was surrounded by electromagnetic fields of significant magnitude, and Jupiter was a source of radio emissions. Lloyd Motz entered the public controversy at that time: though he did not agree with Velikovsky's theories, he felt that Velikovsky should be listened to in view of his erudition and originality, and the success of his predictions could not be gainsaid. In an earlier chapter of this book, Motz provides some interesting insights into Velikovsky's character and into the furor of those days. Also at that time, social scientists entered the controversy, and it is largely owing to their intervention that the Velikovsky affair has become a classic or a cliché, supposedly illustrative of how hidebound and dogmatic mainstream science is, how unwilling to contemplate revolutionary ideas—particularly when proposed by outsiders.

Through the 1960s, the Velikovsky affair was not much in the public eye, but a few academics and students had become fascinated, and in the early 1970s their efforts again led to much publicity: ten issues of the journal *Pensée* dealt with Velikovskian issues, and the American Association for the Advancement of Science held a symposium on "Velikovsky's Challenge to Science."

A full account of these events, an analysis of the controversy from the viewpoint of Velikovsky's scientific conjectures, and references to the copious literature can be found in my recently published book; some reviews of that book offer other viewpoints. The significance of the Velikovsky affair is undisputed, but there is not much agreement as to where that significance lies: as an exemplar of the resistance shown by science and scientists to revolutionary ideas; as an example of public interest in and gullibility toward pseudo-scientific claims; as an illustration of rampant misconceptions about the nature of science; as an object lesson for the scientific community as to how not to engage in public debate. In my view, it is all of those things. I shall, however, argue in this paper that the affair points to the continuing deep malaise of our society that has adopted science as a substitute for religion without yet realizing how unsatisfactory a substitute that has proven to be.

Resistance of Science to Revolutionary New Ideas

It has been claimed that the Velikovsky affair illustrates deficiencies in the way science is carried on, particularly in the way revolutionary new ideas are received: science, it was said, ought to be open to new ideas, no matter their provenance; the pursuit of science ought to be marked by the right to have one's ideas published and tested. To the contrary, I have argued that science is a human and a social enterprise, and that it makes no sense to say that science "ought" to do this or that—science has evolved, it has not been designed, and attempts to control it from the outside have been singularly counterproductive. In any case, science cannot be criticized on the basis of the Velikovsky affair because Velikovsky was not a worker in the enterprise of science and did not play by the rules of the scientific game.[1]

It is true, however, that science is by its nature cautious and conservative, and new notions are not readily taken up;[2] and the Velikovsky controversy does reveal that practicing scientists are not explicitly aware of this.

Contemporary Prevalence of Pseudo-Science

Various individuals and groups, increasingly since the 1950s, have pushed the view that our society is marked by a distressing gullibility toward pseudo-scientific claims. In such discussions, it is common to find Velikovsky's ideas mentioned on a par with astrology, the Bermuda Triangle, the Loch Ness monster, parapsychology, UFOs, and the like. Some have argued that the public receptivity to such matters stems from the loss of traditional religious beliefs; the emergence of new religious cults has been similarly explained.

Misconceptions About Science

I have shown that the Velikovsky affair reveals how misguided are our notions about what science is: Velikovsky and his supporters, the scientists who criticized them, and commentators on the controversy all ascribed to science characteristics that it patently does not possess (or, they argued that science "should" possess those attributes). In that respect, the Velikovsky affair is quite similar to many other public arguments about technical issues—such things as the safety of nuclear power, for example. We are prone to look to science for definite, technical answers to worrisome problems, even when the real issues have to do with judgments and values, not with purely technical matters.[3]

In one way or another, all these points that the Velikovsky affair is supposed to illustrate have something to do with humanity's contemporary approach to knowledge and belief and the part that science plays—or is supposed to play—in that. The belief that science has the answer to all significant questions, and that no answer is true unless it is endorsed by science, has been (pejoratively) described as *scientism;* yet even those who criticize science and scientists reveal through their arguments that they too are scientistic in their beliefs.

This Is a Scientistic Age

At the same time as they claimed science to be empirical, undogmatic, and ever ready to discard outworn theories, scientists who criticized Velikovsky revealed their implicit belief that science is synonymous with certain knowledge, with truth. Thus the astronomer Payne-Gaposchkin lamented that this "scientific" age could lend credence to such erroneous ideas as those of Velikovsky, revealing her implicit belief that a scientific age would be free from error.[4] Two decades later, another astronomer described Velikovsky's ideas as "divorced from scientific reality,"[5] as though "scientific reality"— presumably by contrast to ordinary reality—could be relied on to be true. Not only in the Velikovsky affair does one typically find that scientists implicitly hold scientistic beliefs: when speaking in general or philosophical terms, scientists recognize that science is a continually changing body of knowledge, fallible and with no guarantee of absolute truth; but on specific issues, scientists tend to demand that other opinion should defer to the prevailing scientific consensus.[6]

Commentators on the Velikovsky affair also revealed their implicit scientism, for example by emphasizing Velikovsky's "scientific" evidence and correlations.[7] Just as in advertisements of toothpaste and the like that proffer "scientific" tests, the adjective "scientific" serves in our culture the rhetorical function of ascribing unquestionable validity to the noun that the adjective modifies: correlations, evidence, tests may give us reason to lend provisional credence to something, but *scientific* correlations, evidence, tests tell us that we positively must believe the results.

The argumentation in the Velikovsky affair is replete with criticisms of science and the scientific establishment by Velikovsky and his supporters; yet these criticisms themselves show that Velikovsky and his followers, too, were at root scientistic. In their criticism of scientists and the existing scientific consensus for not recognizing the merit or truth of Velikovsky's propositions, they revealed their belief that science *should and could* encompass all correct investigations of and conclusions about natural phenomena and human events. Thus, they too regarded science as the final arbiter of truth, and their quarrel was only with those temporarily misguided scientists who had unaccountably and reprehensibly failed in their responsibilities to seek the truth wherever it might be found, even in the writings of a man who was not a member of the scientific guild or fraternity. Eric Larrabee, a commentator who strongly supported Velikovsky's right to be taken seriously, stated the point of the controversy to be whether a natural fact could be uncovered independently of science.[8] Now that would be a peculiar question to pose, even only rhetorically, if one were clear about the limitations within which the scientific enterprise works. Larrabee so criticized the scientists who would not listen to Velikovsky *because Larrabee evidently believed implicitly that it is the responsibility of scientists to incorporate into science everything that is true;* which revealed that he too considered science to be the proper repository of all correct knowledge. And Velikovsky himself was as scientistic as anyone. His quarrel was with some details about which contemporary science was supposedly incorrect, and he put forward his own ideas as correcting, *within science,* the outdated notions of such as Newton and Darwin. He did not question; indeed, he too revealed his implicit belief that science is the ultimate arbiter of truth—just so long as science is properly pursued.

Science Has Not Delivered the Goods We Want

The Velikovskians demanded that science expand on the truths that Velikovsky had supposedly discovered,[9] and thereby showed that they misunderstood in yet another crucial way what science actually is. And once again it is not only the Velikovskians but our whole contemporary society that misunderstands in this way. Perhaps the confusion arises somewhat as follows: science is done by human beings and is therefore under human control; science incorporates knowledge of the natural world; the scientific method guarantees freedom from error . . . "therefore" we can use science to do whatever we want.[10]

One flaw in that chain of notions, of course, is the fact that the knowledge gained about the natural world may not be the knowledge that we would prefer to have. Popular discussions of science emphasize knowledge and understanding gained and useful applications resulting therefrom; they do not emphasize that science has revealed some things to be difficult, impossible, or unpalatable. Yet science has shown us limits to the availability of energy and to the degree to which we can safely manipulate the environment, and has given grounds for skepticism that, say, all genetic deficiencies can be overcome. But those limitations on the power of science are not widely recognized, and still less are they understood as reflecting the nature of things. Instead, we hold science responsible as though it *willfully* refused to give us what we want, as though it could give us what we want if only scientists did their job right. We see that in some of the environmentalist and ecologist movements, which ask—just as did the Velikovskians—that science be the handmaiden for their truths, the technical servant that does anything we ask. We see it in the widespread assumption that technological fixes are available for all our discontents. But our deepest disappointment with science, I believe, is not merely that it has turned out not to be user-friendly in technical matters, but that it has failed to provide humanly useful answers to our most pressing questions.

For many centuries, and into the nineteenth century, religion provided answers to the fundamental questions: who are we, why we are here, what we should do. For centuries, religion coexisted comfortably with science, with natural philosophy, which found in the description and understanding of nature nothing but illustrations of God's handiwork. But conflict arose particularly over the matter of man's place in nature, and the empirical power of science eventually carried the day; perhaps in part because the religious authorities would not or could not restrict their dogmatism to moral and spiritual concerns but tried to argue against scientific findings on matters of natural fact. So science vanquished religion as the arbiter of truth within society, and science naturally then came to be looked to for answers to *all* questions, including those having to do with human values and human purpose. But despite efforts in that direction, for example and most recently sociobiology, science has not provided substitutes for the religious parables that taught proper ways for humans to behave. Thus while science continues to be our religion, it disappoints us because it is inevitably an impersonal, epistemic religion and not a personally meaningful moral resource.

The attacks on science that we see, for instance, in the Velikovsky affair are actually a form of anticlericalism: the attack is on the practitioners and interpreters of the religion but not on the religious belief of scientism itself. Velikovsky's challenge to science struck a popular or, perhaps better, a populist chord because Velikovsky appeared to offer a worldview in which science could again explain matters of substantive human concern. He claimed to find true, scientific underpinnings for religious stories; to explain how human consciousness arose in response to natural catastrophes, for instance. In other modern cults of that sort, too, we see how attacks on certain aspects of contemporary scientific views may stem from a desire to find answers to human questions: in the possibility that human existence on earth results from prehistoric visits by superhuman beings, or in the hope that UFOs signify the presence of extraterrestrial intelligences, or in the belief that psychic phenomena reveal untapped energies and capabilities that can mediate impersonal nature and human desires.

Escaping the Dilemma

Such attempts as those of Velikovsky, von Däniken, UFOlogy, parapsychology, and so forth, founder because they find themselves forced to insist on the reality of phenomena that science finds not to be real; just as traditional religions lost their authority when they failed to incorporate what had been learned about natural phenomena. The dilemma of being scientistic cannot be escaped in that direction, by wishfully thinking that the *right* science will somehow give more palatable answers; the dilemma can be escaped only through abandoning the scientistic belief itself: if a supposedly all-powerful science does not have the power to give us palatable explanations for cosmic and human origins, the way out is to recognize that science is not the all-powerful arbiter of truth that we have tried to make of it. The scientific enterprise is, after all, only a human method of certifying knowledge *of a particular sort.* It is uncommonly reliable because of its emphasis on empiricism and consensuality, and it is impressive because of its coherence—the natural sciences have more and more points of connection with one another. Nevertheless, science remains a limited enterprise: it cannot, by its very nature, provide us with a convincing explanation of what happened in "the very beginning." (Steven Weinberg's recent suggestion that science could conceivably at some time discover what happened *before* the Big Bang strikes me as not only quite unrealistic but also as a prime example of scientistic faith.)

If we can become clear about the limited nature of the enterprise of science and yet maintain our respect for its uncommon degree of reliability within its own sphere of applicability, then we can see what sort of religion might be viable nowadays. Like the religions of the past, the new religion should offer parables and myths that promise insight into human origin and nature and behavior; but these must necessarily be accepted as allegorical and not as literal. Religion must not find itself in the situation of trying to contradict science on matters of physics or biology, of trying to substitute its own dogmas for scientific knowledge within science's domain. Rather, religion must stand continually ready to reinterpret and recast its myths and parables so that they can be consistent with—or at least not *in*consistent with—what science progressively discovers.

At the same time, we must beware of extrapolating science in ways that are not valid. Scientism is harmful because it has extrapolated science's materialistic concerns and its reductionist method into claims that only material things are meaningful: for example, that there is no such thing as a soul because it cannot be weighed. But the reductionism of science has been widely misconstrued. Properly construed, reductionism has been useful in providing explanations through progressively more detailed analysis: we can better understand how DNA performs its functions because we know something of the genetic code, because we can understand macromolecular configurations in terms of intermolecular bondings, and so on. But that amounts only to finding more detailed explanations of complex systems by applying our knowledge of some of the properties of some of the "component" parts: it does *not* necessitate a belief that a complex system is *nothing but* an assemblage of its components *behaving just as do those components*. That latter, invalid view is the sort of reductionism that scientistic belief involves, the sort of reductionism that states that a person "is only a collection of chemicals." There is actually no legitimate basis in science for that sort of reductionism. There may well be some physicists who really believe that all properties of an amoeba, say, can in principle be derived from the properties of subatomic particles; but proof for such a proposition is entirely lacking. Rather, science itself offers countless examples that new types of behavior and new types of properties make their appearance as systems become more complex. Thus atoms and molecules "know" nothing about surface tension or other properties of interfaces, still less can we use Schrödinger's equation in meteorology, say. Assemblages of entities give rise to thermodynamic and statistical properties, which have no meaning at lower levels of complexity. So I would hold that it is not fatuous, not scientifically invalid, to hold that human beings have properties of a different sort than are encountered in less complex systems, and moreover properties that we have not begun to fully explore, let alone comprehend. It seems to me clear empirical fact that such concepts as justice and morality are deeply meaningful to human beings, and it is at least possible that those concepts do not arise inevitably from some feature of the wavefunctions of our constituent particles. There is surely room, then, for religious discourse about humanly significant matters, independently of what we have so far learned about pharmacology or physiology.

Religion must not contradict science on matters about which science knows, and science must not claim as its domain areas about which it does not know. Scientific and religious activities are both characteristic of human beings. In certain areas, they overlap. Where that is the case, religion must not try to substitute its dogmas for scientific knowledge; but there is really no reason why religion should wish to do that, since it has literally infinite room to maneuver outside the limited sphere within which science must remain preeminent.

NOTES

1. H. Bauer, "Velikovsky and Social Studies of Science," *4S Review* 2, no. 4 (Winter 1984): 2–8.

2. The best discussion is still the seminal article by B. Barber, "Resistance by Scientists to Scientific Discovery," *Science* 134 (1961): 596–602; see also Bauer, *Beyond Velikovsky*, 295–299.

3. See Bauer, *Beyond Velikovsky*, Chapter 16: "Analogous Cases." The seminal discussion of public confusion of technical questions with questions of values and choices is to be found in A. Weinberg, "Science and Trans-Science," *Minerva* 10 (1972): 109–222.

4. C. Payne-Gaposchkin, "Nonsense, Dr. Velikovsky," *Reporter* 14 (March 1950): 37–40.

5. D. Morrison, *Physics Today*, (Feb. 1972): 72–73.

6. Bauer, *Beyond Velikovsky*, 306–308.

7. For example, J. Lear, "The Heavens Burst," *Collier's* 25 (Feb. 1950): 24, 42–45; L. Stecchini, "The Inconstant Heavens," 80–126, edited by DeGrazia in *The Velikovsky Affair.*

8. E. Larrabee, "Scientists in Collision: Was Velikovsky Right?" *Harper's*, August 1963, 48–55.

9. Bauer, *Beyond Velikovsky*, 200-201.

10. Some such confusion is inevitable when a single word does duty for quite different concepts. The term "science" (or a precursor of it) has long been used to refer to knowledge about nature. Progressively over the last few centuries, the mistakenly oversimplified notion has become widespread that science is safeguarded from error through its use of the "scientific method," and correspondingly "science" has also come to be used as a synonym for "truth."

REFERENCES

Abell, G. and B. Singer, editors. *Science and the Paranormal.* New York: Charles Scribner's Sons, 1981.

Barber, B. "Resistance by Scientists to Scientific Discovery." *Science* 134 (1961): 596–602.

Bauer, H. *Beyond Velikovsky: The History of a Public Controversy.* Urbana & Chicago: University of Illinois Press, 1984.

Bauer, H. "Velikovsky and Social Studies of Science." *4S Review* 2, no. 2 (Winter 1984): 2–8.

Cazeau, C. and S. Scott. *Exploring the Unknown.* New York: Plenum, 1979.

Clark, J. "Velikovsky in Retrospect." *Fate* (June 1985): 97–102.

Cohen, D. *Myths of the Space Age.* New York: Dodd, Mead, 1965.

DeCamp, L. *The Ragged Edge of Science.* Philadelphia: Owlswick, 1980.

DeGrazia, A. *The Velikovsky Affair—The Warfare of Science and Scientism.* New York: University Books, 1966.

Evans, C. *Cults of Unreason.* New York: Delta-Dell, 1975.

Fair, C. *The New Nonsense.* New York: Simon & Schuster, 1974.

Gardner, M. *Fads and Fallacies in the Name of Science.* New York: Dover 1957. First edition was *In the Name of Science.* New York: G. P. Putnam's Sons, 1952.

Gardner M. "Welcome to the Debunking Club." *Skeptical Inquirer* 9 (Summer 1985): 319–322.

Gingerich, O. "On Trans-Scientific Turf." *Nature* 314 (April 25, 1985): 692–693.

Larrabee, E. "Scientists in Collision: Was Velikovsky Right?" *Harper's*, August 1963, 48–55,

Lear, J. "The Heavens Burst." *Collier's*, February 25, 1950, 24, 42–45.

May, J. Review of Henry H. Bauer, *Beyond Velikovsky. Kronos* (forthcoming).

Morrison, D. *Physics Today* (February 1972): 72–73.

Motz, L., chapter in this book, pp. 000–000.

Patterson, J. "Lessons of a Controversy." *Science* 228 (June 1985): 1304–1305.

Payne-Gaposchkin, C. "Nonsense, Dr. Velikovsky." *Reporter*, March 14, 1950, 37–40.

Sladek, J. *The New Apocrypha.* London: Hart-Davis, MacGibbon, 1973.

Smith, R., Review of Henry H. Bauer, *Beyond Velikovsky. Isis* 76 (1985): 428–429.

Stein, G. Review of Henry H. Bauer, *Beyond Velikovsky. American Rationalist* 30, no. 3 (Sept.–Oct. 1985): 53–54.

Stiebing, Jr., W. *Ancient Astronauts, Cosmic Collisions and Other Popular Theories About Man's Past.* Buffalo, New York: Prometheus, 1984.

Weinberg, A. "Science and Trans-Science." *Minerva* 10 (1972): 209–222.

ASTROLOGY AND THE DEVELOPMENT OF ASTRONOMY

8

Classical Astrology:
A Precursor to Astronomy?

KURT STEHLING

Astronomy, the science of the heavens, began in ancient times primarily in response to the needs of organized agrarian societies, in order to understand and predict the seasons that regulated and controlled harvests, flooding (as in Egypt), and the like.

In many ancient theocratic societies, as in Mesopotamia, China, and the "New World," the dominant priestly classes soon associated celestial events—such as lunar and solar eclipses, planetary and lunar positions and movements, comets, and even meteor showers—with important religious festivals and practices.

Thus began the empirical practice of connecting celestial happenings and positions with human events and conduct, and even personalities, aside from political and religious matters. Astrological practice became more widespread, believable, and systematic, and has survived and even expanded into our own times.

Although it is difficult to untangle early astronomical and astrological practices, there seem to have been mutually beneficial interactions between them from their very beginnings. While hieroglyphics and Sumerian cuneiform tablets, for example, have shown that astronomy (as a science) influenced astrology, there is less basis for the reverse.

Nevertheless, a case can be made for astrological methods and beliefs influencing the science of astronomy and astronomers up to, and including, Kepler.

The Alchemy Analogue

Before pursuing the astrology-astronomy relationship, it may be of interest to compare the influence of another empirical "science"—

alchemy—upon chemistry. Alchemists began their chimerical quests for the Elixir of Youth, the transmutation of base metals into gold and silver, and the discovery of a universal cure for disease in medieval times—long before the science of chemistry was established in the late eighteenth century.

While the alchemists did not, of course, achieve their goals, they (perhaps unwittingly) set the stage for the development of chemical science. Among other apparatus, they bequeathed to chemists such items as retorts, kilns, muffle furnaces, glassware, crucibles, tongs and other handling gear, test tubes and glass flasks, high temperature combustors, fume hoods, and so forth.

Of equal importance, the alchemists left a legacy of techniques that the budding chemistry community of the late eigteenth century adopted and adapted. These include color indicators for bases and acids, reduction and oxidation processes (e.g., cinnabar reduction to mercury and then the metal's precipitation and amalgamation—important processes for the hoped-for mercury-to-silver transmutation), distillation, alloying, grinding and filtering, smelting and flotation, and various reagents. Lavoisier and his colleagues, and eighteenth and nineteenth century transition successors, noted—albeit indirectly—their debts to the disappearing alchemists.

On the other hand, the influence of astrology upon astronomy is much more diffuse and ambiguous and no short transition period (as in alchemy-to-chemistry, from, ca. 1750–1800) can be identified.

Ancient Relationships

Some facts are known about Chaldean, Egyptian, and Greek astrology's relationship to astronomy. Something is known about *each* subject (as in Chaldean cuneiform horoscope tablets) and, of course, from such early writings as the *Almagest* of Claudius Ptolemy or his *Tetrabiblos*. The latter is most concerned with astrology coupled with astronomical references and symbolism.

The Ptolemaic writings and theories show indubitably, as do those of other early philosophers, that mysticism, a belief in the occult, and interests in empirical philosophies were intertwined with their "other side," that is, a belief in objective and rational disciplines—such as astronomy. It must be remembered that the ancient—even the early medieval—worldview was much more circumscribed, even for the most astute and iconoclastic thinker.

Chaldean records, as early as 2000 B.C., have shown that Chaldean astrologers (actually an amalgam of astronomers and astrologers) had already found and recognized the following:

- The obliquity of the ecliptic
- Some notion of the motions of the five visible planets, i.e. Mercury, Venus, Mars, Jupiter, and Saturn
- The permanence of many star groups, i.e., constellations
- Meteors as "heavenly visitors," but not "falling stars."

Although the Chaldeans ascribed certain "vapors" or influences on human (i.e., royal) affairs, along with these supernatural manifestations, they also recognized the purely astronomical implications and recorded these phenomena without astrological insinuations.

Western Hemisphere Astrology

On the other side of the world, in the Hawaiian islands to be precise, astrology and astronomy were conjunctive, although the latter was never as well and precisely recorded as the observations of the ancient civilizations several thousands of years earlier. There seems to have been no particular distinction made between astrologers and astronomers; the priestly groups used heavenly phenomena—as did their confreres in other societies—for influencing their superiors, the kings. King Kamehameha II had a staff of about forty priest-observers who kept him apprised of the astrological implications of heavenly apparitions—comets were celestial tours-de-force for the priests—but they also seemed to be aware of celestial influences on such phenomena as tides. The first European visitors to Hawaii during the eighteenth century were impressed by the astronomical erudition of the native priests.

Captain J. Dutton of the U.S. Army during a seven-month period in Hawaii during 1882, stated that the priests' astrological knowledge and records (in the form of pictographs, there being no formal writing) of celestial events such as eclipses rivaled those of more advanced societies.

Yet another society, the Mayans, also had an elaborate framework of celestial comprehension. Their observatories, such as the "Caracol" at Chichen Itza in Yucatan, were obviously designed to permit unhampered scanning of the clear Yucatan skies by the priests-cum-astrologers/astronomers. They used their observations and predictions for influencing the Mayan ruler, and for impressing the populace.

The remaining ruins of the Caracol are covered with mystical (astrological?) carvings and reliefs, which may also have been designed to punctuate the dates of the vernal equinox, the autumnal equinox, and the winter solstice. Here, then, we note a complementary mix of astrological and astronomical conjunctions, along with actual measurements and observations.

Arabian Astrology

To the ancients, heliocentricity, the "New World," the relationships and extent of the ocean, atmosphere, and earth, the nature of the stars and planets, human physiology, and "scientific" medicine were unknown. Therefore, to allow the intellectuals of their time to come to grips with, or understand, their world and the behavior and organization of nature, and even their own bodies, it was necessary for them to "hedge their bets" with beliefs in supernatural forces, medical mysticism, and such diversions as palmistry and, of course, astrology.

Astronomy, mathematics, and astrology flourished in the Arabian world from the eighth century on. Physicians, for example, had to know astrological portents to find the most favorable times for medical treatments.

In Baghdad, during the eighth and ninth centuries, there flourished an ordered and comprehensive school of astronomy and astrology, with the two elements intertwined and mutually influential. This school of scholars drew heavily upon Judaic and Christian experiences and literature. Astrology was considered a conjugate of astronomy; the reverse could also be claimed. One of the more important astrological works to appear during this period was written by Jafar Abu Ma'shar or "Albumazar."

This book, considered a standard work during the Middle Ages, dealt with the positions of the planets and signs and explained their meanings. The book was also used as an astronomical reference source by later astronomers, such as al-Battani (or "Albategnius"), in the early tenth century. Battani, who published his own rigorous astronomical text, gave credit to Albumazar for such ideas as the use of armillas, quadrants and even the astrolabe—which may have been used by his predecessors (including Albu) for the positioning—and predicting—of heavenly bodies for astrology.

There is little doubt, then, that during the eighth, ninth, and tenth centuries of "Arabian"* scientific enlightenment, astrologers and astronomers were often the same men and, when not, the disciplines interacted; with astrology—a better funded endeavor by the Caliphs—being dominant over the teachings and practices of astronomy.

Arabian astrology and astronomy spread far and wide. For example, the writings of the above scholars, among others, appeared (not unsurprisingly) in Spain. There, for example, the Castilian King, Alfonso X, or "Alfonso the Wise," surrounded himself with astronomers to produce new tables of astronomy. The leader of this group was the great Jewish scholar, Isaac ben Said. He acknowledged the debt owed by him and his colleagues to Arabian astrology and, of course, astronomy. He was a friend of the King's astrologer (an ancillary and honored member of the group) and was a frequent visitor to the extensive library of another Jewish scholar, Abram al-Wefa.

Asian Astrology

Chinese astrology and astronomy were well developed and continuing endeavors until even recent times, with a resurgence of astronomy and astrophysics now under way.

---

*"Arabian" here really means Jews, Syrians, Persians, and other peoples of the Levant and Mesopotamia.

The need for reckoning time motivated the early Chinese to expand their knowledge of astronomy. During the Chou period, a century rich in literature and philosophy, the movement of the constellations and planets were charted and predicted, along with the seasons.

The astronomical activities of that time were actually "astrological" in that the "meaning" of the various phenomena were discussed in terms of their effect on human events. The court astrologer was also the court astronomer. The astronomical scholars were supported by the emperor and the state since they also prepared horoscopes and described the sky and its bodies in such terms as, "the North Pole and its surroundings, at the peak of the heavens," was the seat of the emperor. The philosopher Shih-shen, buttressed by the emperor's interest in, and dependence upon, astrology, wrote an astrological work of which many fragments survived into later times. His astrological work was really an astronomical catalogue containing several hundred entries of constellations and stars.

Although he drew on earlier works, much of the material was new, based on his own studies and observations. Again, the point is that astrological "data" and the supporters of astrology provided the basis and support for astronomical work—although at that time it was not called that. This catalogue, by the way, was earlier than the time of Hipparchus, ca. 130 B.C. Later Chinese scholars, such as Shuging (1231–1316), were, one might say, bread-and-butter astrologers, able to blend their belief in empirical practices with development of astronomical mathematics, such as those used to improve the tables of the sun and moon. The Mongol Emperor Kublai Khan provided funds for such scholars to build shadow poles, quadrants, and armillas, provided that they "kept him abreast of astrological forecasts."

Medieval, Renaissance, and Enlightenment Activities

Returning to medieval Europe, we find that while astronomy began to flourish (after the "Dark Ages" hiatus), astrology was doing very well indeed and was to provide the underpinnings for much astronomical thought during the twelfth to fourteenth centuries. The astronomical worldview, as proposed, for example, by John of Holywood (d. Paris 1256), was couched in astrological terms or "descriptors" which, nevertheless, had a pragmatic underpinning. His writings could, in fact, be used by a student of astronomy, since the actual and predicted motions and appearances of heavenly bodies were listed. Similarly, an "astrologer" could read what he wished into the text, diagrams, and tables to connect such heavenly bodies' motions and appearances with worldly goings-on, such as floods, deaths (or births) of Kings and Princes, and so forth.

Dante's *Divine Comedy*, in which he described the Celestial Sphere, heaven and hell, and Lucifer at the center of the earth, also had astrological implications. His views reflected the "Weltanschauung" of those days (thirteenth century) which was not confined to the earnestly thoughtful structure of Aristotle's cosmos, but instead, was dominated by astrology. This appeared as the encompassing doctrine of the world. Medieval thinkers, including Albertus Magnus (1193–1280) and Thomas Aquinas (1225–1274) believed that the stars' movements dominated life on earth. This belief was, for them, consistent with their more objective and deductive speculations about the nature of the cosmos, particularly the place of the earth within the celestial sphere.

During the Renaissance, speculation and inductive reasoning were superseded by deduction, observation, and experimentation. Celestial navigation became an established science whose tenets were derived from Arabian science and such Jewish astronomers as Jacub Carsono, whose patron was King Pedro III of Catalonia. Celestial navigation not only permitted long ocean voyages leading to the discoveries by the Portuguese and Spaniards, but created a fruitful feedback into the consequently expanding science of astronomy.

One might think that with astronomy becoming a viable and flourishing science with practical applications, astrology would wither and become inconsequential. But instead, belief and interest in astrology became even more widespread in the sixteenth century; it was more widely studied, or at least accepted, than the pragmatic activities of astronomy. Almanacs, horoscopes, and calendar sheets were widely distributed. They contained not only celestial phenomena, such as eclipses and conjunctions, but also weather predictions and medical advice.

The princes of Europe employed, as did their predecessors, court astrologers who were called "mathematicians." It might seem that such astrological diversions would detract from the resources—financial and otherwise—needed by the growing body of observational astronomers. Instead, the opposite occurred, at least for the "father of observational astronomy," Tycho Brahe (1546–1601).

This Danish nobleman had, as a youth, shown not only a great passion for observational astronomy, but also for astrology. He believed that, by careful observation and study, astrology could move into the undisputed ranks of rigorous science. In a lecture at the University of Copenhagen in 1574, he stated that astrology was the chief practical objective of mathematical discipline.

It was his belief in the astrological doctrine of the unity of heaven and earth that brought him into scientific research, that is, mainly systematic and accurate astronomical observations and measurements. His observations of a new star, a nova—or rather a supernova—in November 1572, expanded his zeal for astronomy; yet he speculated not on its origin or dynamics, but rather on its astrological implications as to what great events it could foretell.

In 1575 he visited Cassel on his travels through Germany, where he gained the attention of the Landgrave Wilhelm IV of Hesse, who was in the process of establishing an astronomical observatory. Wilhelm was so impressed by Tycho's erudition and enterprise that he wrote his friend King Frederick of Denmark a laudatory letter wherein he suggested that Tycho would become a famous Danish astronomer if he were given the necessary patronage by King Frederick. The King had earlier given Tycho some support for his—mostly chemical—studies. More importantly, he had long been interested in astronomy and astrology. Although apparently no great believer in horoscopes, he was fascinated by the astrological writings and predictions of the ancients. As a youth his interest had been aroused on this subject by one of his tutors who showed him star charts and tables illustrated in the many extant treatises on astrology.

Thus, the king needed little encouragement from Wilhelm to urge Tycho to return to Denmark, where he established him subsequently on a small island near Copenhagen. Tycho then built his world-famous observatory, "Uraniborg."

At any rate, had Frederick not been imbued with the romance of astrology that started him on his heavenly interests, it is doubtful that he would have diverted his moderate resources (against opposition from some of his nobles) to Tycho's great observatory. And to extend the linkage, Kepler, Tycho's trainee, might not have got his chance (later, in Prague) to break the back of the anti-Copernican opposition. Kepler's genius accomplished this, but in this achievement, he depended on the enormously detailed and accurate astronomical data accrued during Tycho's twenty years of painstaking observations at Uraniborg. During that time, Frederick spent many hours discussing astrology with Tycho, who never lost his interest in that subject.

Kepler himself had absorbed many astrological ideas from Tycho; he treated astrology not as a mystical system of conjectures but rather as a doctrine of world unity. His astronomical insights arose—by his own admission—because his astrological interest motivated him to probe and understand the secrets of the heavens. It need hardly be said that he succeeded very well indeed!

In conclusion, then, it can be said that astrology played an important historical role in influencing the development of astronomy, at least until the early seventeenth century. While many charlatans were later to dominate astrological speculations, the great scholars of antiquity, the so-called Dark Ages, medieval times, the Renaissance and then the scientific "Transition Period" of the sixteenth and seventeenth centuries were comfortable with the subject and, as noted before, used it as inspiration for the deductive science of astronomy.

9

Why Does Astrology Still Flourish?

CHARLES R. TOLBERT

In ancient times the activities of the astrologer and of the astronomer were not seen as separate functions, and could only be distinguished by the activity engaged in. The astronomer/astrologer was generally the same individual. The same duality applies to the fields of alchemy and chemistry. A parallel between the two is made. Because of their understanding of nature, the ancients perceived astrology (alchemy) to be useful to the practitioners and to the society. This perception stimulated study of the field which eventually led in part to the separate field of astronomy (chemistry).

Unlike alchemy, astrology still persists as an active practice today. Why the continued popularity of this discredited field of study? Some reasons are ignorance, wish for luck, explanation about the unknown, low investment of time and resources, failure of science to fully disprove it, and the human tendency to remember successes and to forget failures. Astrology may have been a rational activity for the natural philosopher of the past, given his knowledge of nature. It is not today. It has been replaced by "the laws of science." Will the intellectual of the future look back on our activities and still see them as valid or will he have the same view of us as we have of the ancient astrologers? Each era sees its science as correct and finds it difficult to foresee the science of the future.

Kurt Stehling reviews the early history of astrology and compares its influence on astronomy with the early influence of alchemy on chemistry. He concludes that "astrology played an important historical role in influencing the development of astronomy, at least until the early seventeenth century."

Stehling restricts himself to the comparison between astrology and astronomy prior to the time of Kepler. During these early times there was little distinction to be made between the two intellectual pursuits. As Stehling points out, frequently the astronomer of the day is also the recognized astrologer. The individuals themselves seldom made the distinction. Ptolemy, the last great Greek philosopher of antiquity, gave us the summary of Greek astronomy in *The Almagest* and the foundations of astrology in *The Tetrabiblos*. A similar question can be asked with respect to chemistry and alchemy. Were the alchemists and the chemists different people or was it essentially the activity they engaged in that defined their character?

If the astrologer and the astronomer were the same individual, then the astrological interest, along with the interest in floods, eclipses, and seasons, was just another of the motivating forces in studying the heavens.

It was necessary to study the positions of the planets, the sun, and the moon in order to predict the seasons for planting and harvesting, to predict the flooding of the Nile, and the other natural phenomena. To extend this prediction mode to events that affected individuals or nations would seem natural to the intellectual of two thousand years ago. This desire to predict would fuel a desire for accuracy. The need for accuracy certainly stimulated observation of the heavenly bodies and the further understanding of the theory regarding their behavior. There is no question that astrology motivated the ancients to improve their understanding of nature, hence was perceived to be useful.

Similarly, alchemy stimulated chemistry. Here it was the perception of potential wealth that led the alchemists to struggle with nature in their attempts to use the fledgling science of chemistry to produce precious metal from baser material. It is easy even for us today to understand this motivation. Had they proven successful, wealth and fame would have been theirs. As more and more was learned about what happens when one attempts to alter the state of a material chemically, it became clear to the alchemists that their science, or pseudo-science, was not likely to achieve its goals. It was, in fact, developing other insights into nature that proved just as useful to mankind. There gradually grew a divergence between the people who felt the goal of turning a base metal into precious metal was still achievable and those who began to use alchemy for other benefits to mankind: this latter group gradually became the chemists. The alchemists, if you will, died out of the system but the chemistry of today owes much of its early knowledge to the attempts by the alchemists to better understand nature for their peculiar needs. It is at this point that the comparison between alchemy and astrology fails.

After Kepler and after Newton, mankind began to realize that the heavenly bodies were not controlled by gods and were not there for the impact they might have on human lives. There gradually arose a divergence between astrologers who still felt that the positions of the planets influenced human nature and the astronomers who studied the heavens to aid their understanding of the laws of nature. The early knowledge of the behavior of the planets and stars grew through the efforts of people whose goals were partially astrological. Yet as it became clearer that astrology had no real basis or hope of success, the study of the sky was taken over by astronomers. However, there remained an active and viable group of astrologers still pursuing the original astrological goals. This was simply not the case in the alchemy-chemistry situation.

What is it about astrology that maintains its viability in the face of scientific understanding that it cannot and does not work? Why is astrology in this regard different from alchemy? Once it became clear that you could not turn base metal into gold the alchemists died out and left the field to the chemists. I am sure most of the astronomers of today wish a similar thing had happened with astrology. Why does the discredited field of astrology continue with such persistence in the modern world? Why have the twins of astrology-astronomy deviated in their pattern of development from the counterpart twins alchemy-chemistry?

One interesting clue is available if we read Stehling carefully. He points out that astrology was the predictive icon of the upper and ruling priestly castes and classes. Originally astrology was something carried out by and done for the intellectual elite. Today astrology is generally a more firmly held belief by the less educated and less well-off individuals. As has sometimes been said of religion, astrology has become a palliative for the masses. Human beings need to have someone to blame or bless for the events that occur to them. Why has the crop failed or been lush? Who causes the floods and the earthquakes? The events beyond our control, and sometimes beyond our understanding, require a reason or cause that we can relate to ourselves. In so uncertain a nature, how can we make decisions about our actions? Astrology (and religion) can provide these answers and soothe our troubled psyche. This is the explanation of the continued "usefulness" of astrology.

In addition, there is simple cost-benefit analysis. In the case of alchemy it turned out that while you can get a considerable benefit for your investment of time and energy by doing chemistry, the chance of turning base metals into gold is so low that it is not worth the time and effort involved. In the case of astrology, on the other hand, there is very little time and effort involved (except for the professional astrologers, and they find great wealth in the field). For the rest of us, dabbling in astrology is like carrying a lucky rabbit's foot as an omen of good luck which costs us very little and *might* have some beneficial effects.

As the science of chemistry progressed in its understanding of nature, it became clear that it was impossible to turn base metal into gold. No such impossibility confronts the astrologer in the science of astronomy. As the astrologer is quick to point out, "the stars impel, they don't compel." And there is that blessed human characteristic that forgets the negative and remembers the positive. How many times have we heard of "the miracle" that saved someone from the air crash that killed all of his fellow travelers. We forget that the same source of "the miracle" was also the source of the original disaster. Just as the gods are only remembered for their blessings, so is astrology only remembered for it successes. Whether we view astrology as a kind of religious belief, or as a cocktail hour pastime, it is clear that today its basis is not scientific and its continued presence in the modern world is an anomaly.

The interest of human beings in the potential influence of heavenly bodies is egocentric. We consider ourselves—as individuals, as nations, as races, or as a species—distinct from other animals and special in this world. For much of early history, man assumed that the world was set apart for him and that nature operated to his benefit. This strong trend of human thought led us, to believe that things that happen in nature must be happening with some regard for us. Astrology ascribes these happenings to the effects of the planets and the stars on us.

The scientists of today find it easy to say that astrology is "bunk." They know that our lives are controlled by things like the law of gravity, Maxwell's equations, and the theories of atomic and molecular chemistry. The ancient astrologer knew that there were major events occurring in his life and in the life of his nation—events that made him happy or sad, peaceful or warlike—and he ascribed their occurrence to the planets and the stars. Today we also realize that we are happy or sad, warlike or peaceful, but instead ascribe these influences to "the laws of nature." Will some future scientist look back on our belief in the influences of the laws of nature on human behavior and refer to it as "bunk"? Each era sees its science as correct and finds it difficult to foresee the science of the future.

Is it not normal human behavior to look to nature for connections and generalizations that will be useful in making judgments in our lives? The ancient astrologers were intellectuals just as the scientists of today and their choice of astrology as a paradigm for dealing with human behavior was not hard to understand. It is the modern astrologers and their public acceptance that leaves us baffled.

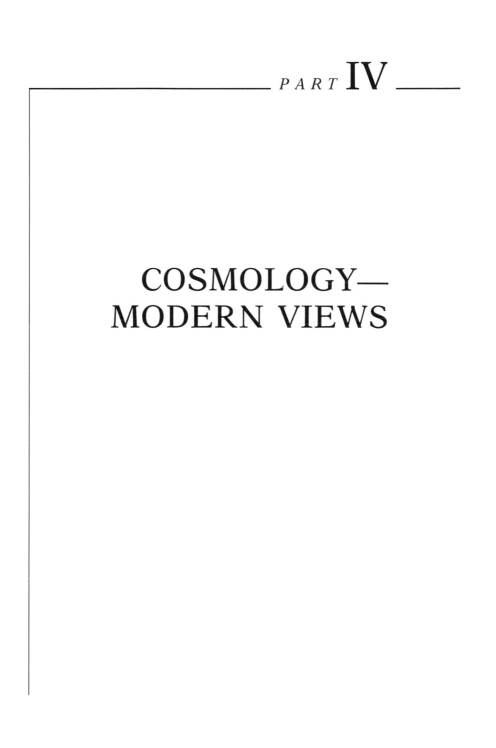

PART **IV**

COSMOLOGY—
MODERN VIEWS

10

Cosmology—Modern Views

HONG-YEE CHIU

Tracing the road back to the time of Copernicus, development of modern cosmology started when the first measurements of stellar distances were made in the 1800s. During the early twentieth century the modern view took shape. First the discovery of the so-called extragalactic nebulae enlarged our views beyond our Milky Way. Next Einstein came up with the theory of general relativity, which enables us to obtain a large-scale structural model of our universe. Work of Friedmann and Lemaître, based on Einstein's theory, proved theoretically that the universe must be dynamic, that is, it must be either expanding or contracting. The observational work of Hubble proved that, indeed, our universe is expanding, thus confirming the cosmological model of Lemaître and Friedmann. Gamow and Alpher created a physical cosmological model—the big bang universe—based on the work of Lemaître and Friedmann. The big bang model became the foundation of modern cosmology. In the 1980s emphasis was placed on the first moment of creation—which may explain many otherwise inexplicable observed phenomena, such as the origin of homogeneity and isotropy in our universe. The inflationary model is one such model being pursued today.

Introduction

Cosmology is the study of the origin, constitution, structure, and evolution of the universe. As we know today, the universe extends far beyond the Solar System. Much of our current knowledge of the universe, out to perhaps 10^{10} light years, is derived from data collected by giant telescopes. At such distances, even the brightest galaxies appear as specks of light. To decipher the messages brought to us by light that has traveled billions of years, we need not only the most sophisticated instrumentation to detect and sort out the signals, but also theoretical knowledge to interpret the data collected. The edge of the universe is far beyond the limit of vision of our largest telescopes. Indeed, progress in cosmology depends on a combination of theoretical and experimental efforts.

Since the dawn of civilization, man has never given up searching for the unknown, and one of his major tools is reasoning. From observation of his surroundings, aided by experiments designed by him, man gradually enlarged his tools of reasoning, and thereby his knowledge of the universe. Although in all likelihood we will never be able to leave our own galaxy, the Milky Way, to cross the gap of interstellar void to visit even the closest galaxy, for example, one of the two Magellanic clouds, we understand other galaxies as if they were our own.

Can the same physics laws derived from observations carried out on our earth be used to interpret data from the most remote part of our universe? As technology advanced, the degree of precision with which our physics laws are confirmed in earth laboratories increased so that there is no doubt as to the applicability of these physics laws even in the remotest corner of the universe. Thus, we can proceed with the assumption that all physics laws observed on our earth are applicable to cosmological problems.

Historical Background

We can describe advances in our views of our universe in two respects: observationally and theoretically. Although advances in these two aspects seemed at times to be disparate and unrelated, they eventually merged (especially within the last fifty years).

The history of cosmology has been one of continued diminution of man's position in the universe. Nicolaus Copernicus (1473–1543) was the first in the Western world to espouse the view that the earth is not at the center of the universe, thus enlarging our view of the universe from our earth to the entire Solar System. A century later, Giordano Bruno (1548–1600) enlarged our view even further. He postulated that all stars were like our sun and that the reason for their faintness was that they were far away. (In present terminology, Bruno enlarged our universe to a size of tens of light years).* Sad to say, one believer of Copernicus' theory, Galileo Galilei (1564–1642), paid for his belief with his freedom, and Bruno, with his own life. Since then persecutions for holding nonestablishment cosmological views have virtually stopped.

Observational View

Advances in telescopes and associated mechanical improvements, notably in the nineteenth century, further enlarged our views

---

*One light year is nearly 10^{13}km, or 10,000,000,000,000 km.

of the universe to tens of thousands of light years. For the first time, the distances of some nearby stars were actually measured. It was then recognized that our Milky Way was a conglomerate of stars like our sun, some brighter and some dimmer, as Bruno had predicted.

The twentieth century brought major technological developments. The use of electricity became widespread; results of the industrial revolution were felt around the world. Man was imbued with extreme confidence in his abilities, as witness the "unsinkable" Titanic. That tide led to the then biggest astronomical instrument, the Mt. Wilson 100-inch telescope that brought surprising discoveries to enlarge our view to hundreds of millions of light years, laying the foundations of our current cosmological theories.

The 100-inch telescope was powerful enough to resolve individual stars in some of the nearby nebulae, now called galaxies. There is no longer any doubt that these nebulae are indeed star systems similar to our Milky Way. Further, a number of a type of variable stars, called Cepheid variables, were discovered within these galaxies. These stars were known to have a precise relationship between their brightness and their period of variation, ascertained through the laborious efforts of Henrietta Swan Leavitt and Harlow Shapley. Knowing their real brightness, one could establish the distance scale to the nearby galaxies: the distance to the Andromeda galaxy, the brightest galaxy in the sky other than our own, was around 500,000 light years. (In the 1950s this distance figure was revised to over 1,000,000 light years.)

Vladimir M. Slipher, an astronomer using the new 100-inch telescope between 1895 and 1914, successfully measured the Doppler shift of spectral lines of a number of nebulae, thus establishing their line-of-sight velocities. Velocities of stars in the Milky Way had been measured earlier. Typical values of stellar velocities were in the tens to at most 100 kilometers per seconds (km/s), with some stars moving towards us and others moving away from us. However, as Slipher discovered, the velocities of certain nebulae were all negative (moving away from us) and were unusually large, ranging from a few hundred to 1,000 km/s.

Edwin Hubble combined Slipher's observations of the velocities with the distances to the galaxies deduced by Leavitt and Shapley. Adding his own observations up to 1929, on galaxies several hundred million light years away, he discovered that there was a definite relationship between distance and velocity of recession of the galaxies. He thus showed that the universe was expanding, a view enhanced by all later observations.

Theoretical View

Although there had been many efforts to formulate a theory of the universe throughout history, Sir Isaac Newton (1642–1727) should be credited as the first who successfully unified observations of the visible universe at that time, the sun, moon and planets, into a theory of mechanics and gravitation still valid today for most applications. His theory explains the motion of planets via the concept of gravitation. Despite its success in the prediction of the orbit (and of the return) of Halley's comet, and the prediction of a new planet (Neptune), a cloud hangs over some of its predicted consequences. At the end of the eighteenth century Marquis Pierre Simon Laplace (1749–1827) studied a hypothetical large mass, and he concluded that "A luminous star, of the same density as the earth, and whose diameter should be 250 times larger than that of the sun, would not, in consequence of its attraction, allow any of its rays to arrive at us; it is therefore possible that the largest luminous bodies in the universe may, through this cause, be invisible." Here he touched a most interesting subject, beyond the scope of Newton's original theory of gravitation. Laplace innocently crossed the boundary of Newtonian mechanics into Einsteinian mechanics (general relativity), a theory that was not to be developed for another century.

No one has yet formulated a theory of the universe as successful in its predictions as Einstein's. Yet his 1915 theory of general relativity, at the time it was developed, could only predict a dynamic universe, one in which the entire universe is either in a state of expansion or contraction. (The expansion of the universe was not discovered and confirmed until the late 1920s.) Right after the introduction of his general theory of relativity, Einstein tried unsuccessfully to generate a static model of the universe by means of his theory. In 1922 the Russian mathematician Alexander Friedmann derived a model of an expanding universe, in strict accordance with Einstein's theory. Unfortunately Friedmann's model was not known to many (his country being in a state of turmoil and revolution) and a Belgian priest, Abbe G. Lemaître rediscovered it in 1927. Again, Lemaître's work was published in a rather inaccessible journal, and in the words of another great astronomer, Sir Arthur S. Eddington, "seems to have remained unknown until 1930 when attention was called to it by de Sitter and myself." In the theories of Friedmann and Lemaître, the universe began from a singular point, expanded either forever, or to a maximum extent and then contracted to a singular point again. Although this theory has theological overtones and was attacked as such, Lemaître's model was based on strict scientific deductions without any ideology.

In the interim, a number of other theories flourished. The most notable one was the steady-state universe, proposed in a sequence of papers starting 1948, by Sir Fred Hoyle, Thomas Gold, Herman Bondi, and R. A. Lyttleton. A static-state universe has the beauty that it has always been there. On the other hand, the expanding universe appears to be a fact of life. The steady-state universe combines the static-state universe with the expansion feature. In an expanding universe, matter density always decreases. In order to maintain a steady state, however, one is forced to postulate spontaneous creation of matter. Although this theory had many attractive features, one of its weaknesses was in the difficulty in the interpretation of the background microwave radiation discovered in 1963. (See below.)

Now it may seem pedantic to dispute the correctness of the idea of an expanding universe, a theory now taught even in elementary schools. However, it must be remembered that by the time the expansion of the universe was predicted and confirmed with observations in the late 1920s, quantum physics was still in its infancy and not completely developed. The neutron, a constituent of the nucleus, was not to be discovered until a few years later, and the nuclear source of stellar energy was not established until 1938 when Hans Bethe and C. L. Critchfield published the first precise formulation. Many microscopic properties of matter were not known. By the late 1940s, nuclear physics was under rapid development. At this time, George Gamow and his student Ralph Alpher decided to add physics to the Friedmann-Lemaître model. This combined theory is now known as the "big bang" theory, since the universe appeared to have been created amidst an explosion process.

The Friedmann-Lemaître model is a simple model, consisting of uniformly distributed matter and radiation. We know there is interaction within matter itself and between matter and radiation. These interactions produce stars, galaxies, and other objects. Gamow and Alpher described the interaction of matter and radiation within the framework of the Friedmann-Lemaître model. In fact, many of their conclusions remain valid to this day. For example, their theory predicted the existence of a general, extremely cold background microwave radiation, with a temperature of a few degrees Kelvin (°K) above the absolute zero, as well as the existence of primordial helium, synthesized within the first few minutes after the creation of the universe.

The past decade has seen new developments in cosmology, as our knowledge in physics advanced. Accepting the basic concepts of the big bang cosmology, theory has focused on understanding the creation process and the subsequent evolution, when density and temperature are so great that conventional theories of matter are no longer applicable. The result is the so-called "inflationary model," according to which, during the very early epoch, the universe went through an inflationary phase whereby it achieved many of the large-scale properties we see today, such as isotropy and homogeneity.

Current Knowledge of the Universe

Following this brief historical survey I wish to give a more organized review of various properties of the universe, in conjunction with theoretical developments. The discussion will be divided into three parts: observable matter, radiation, and invisible matter.

Observable Matter of the Universe

As stars are building blocks of a galaxy, galaxies must be regarded as the fundamental building blocks of the universe. When we look outside our galaxy, the Milky Way, the universe is largely void and the only visible occupants are galaxies and a small number of other entities such as quasars. So far, at least, any attempt to discover matter between galaxies has been unsuccessful. We can safely assume that there is no intergalactic matter.

Galaxies are made of gas and stars, in a ratio of roughly one to ten (one part gas by mass to ten parts stars). The age of stars varies, for known stars, from a few hundred thousand years to as old as 20 billion years. The composition of stellar matter also varies. Since all chemical elements are synthesized from the lightest element, hydrogen, younger stars tend to contain more of the heavier elements than old stars, a fact that is confirmed in almost all instances. An exception is helium, which, according to the big bang theory, was synthesized shortly after the creation of the universe, with a concentration of around 20 percent. It is expected that even in the oldest stars the helium content would not fall below this amount. But because helium is very difficult to detect—at least in old stars that have lower surface temperatures—this point has not been satisfactorily settled.

The distribution of galaxies in the universe appears to be fairly uniform, with a degree of clustering that can be explained in terms of random fluctuations. However, the importance of clustering probably has been underestimated in the past. In addition, latest observations suggest that the largest cosmic features may assume the form of filaments, surrounding empty blobs. All these problems probably will become better understood in the near future upon the completion of the Large Space Telescope, now scheduled for launch in 1990 at the earliest. In short, it can be said that, subject to the uncertainties discussed above, the distribution of galaxies is generally uniform in all directions (isotropy) and in all locations (homogeneity).

From data of all galaxies discovered so far, a uniform state of expansion of the universe is observed. The state of expansion can be expressed in terms of a linear law, known as Hubble's law. The rate of expansion is 75 km/s per megaparsec of distance (with a probable error of ±50 percent). That is, at a distance of 1 megaparsec (3.26 million light years) the expansion velocity is 75 km/s, and at a distance of 2 megaparsec the expansion velocity is 150 km/s, and so on. According to this equation, at a distance of 4,000 megaparsec, the expansion velocity would be the velocity of light. Hubble's law thus must be modified in order to take account of relativistic effects near that distance (say, at distances greater than 2,000 megaparsecs).

In addition, when we look at distant galaxies, we are also looking at the past—thus looking at younger galaxies. Therefore, looking at galaxies at 1,000 megaparsecs away, we are looking at light emitted more than 3 billion years ago. Things of course can be very diffeent if we look far back enough.

Quasars are objects that exhibit very large red shifts—being presumably very far away, and thus young. They appear to be massive objects of rather small size, emitting large amounts of energy with violent activity. Because they have large red shifts, they may represent what went on during the very early stage of our universe. Although there have been many theoretical studies about the nature of quasars, none of them seems to present any consistent answer. Most likely they are galaxies in the earliest evolutionary phase. However, the inconsistency in their properties led to the belief that they are much more complicated objects than galaxies. Even the applicability of Hubble's law to correlate their distance and their red shifts has been questioned. We should regard quasars as one of those remaining mysteries to be solved in the near future.

The density of matter in the universe can best be expressed in terms of number of hydrogen atoms per cubic meter (h/cbm). A current view, presented by Geoffrey Burbidge, is that the density of matter due to galaxies is 0.07 h/cbm. The most aggressive estimate cannot exceed 0.2 h/cbm (largely due to the uncertainty in the value of the Hubble's constant discussed above). As we will see later, it requires a density of matter of at least 3 h/cbm to "close" our universe—to make it contract by self-gravitation—a topic that we will cover later in this article.

Radiation in Our Universe

In 1964 Arno A. Penzias and Robert W. Wilson discovered that there is a uniform background microwave radiation (i.e., coming from all directions) in the universe, with a radiation temperature of $3°$ K. This radiation had been predicted in the big bang model some fifteen years earlier. Since then, much research has been carried out to study details of this radiation. The discovery of this radiation, in addition to lending a firm support to the big bang theory, is important in the following respects:

1. It establishes an absolute stationary coordinate system in our neighborhood. Since the background radiation must also be subject to red shift, it can be isotropic only in one frame of reference; only when the observer is not moving with respect to the center of gravity of the expanding universe, will he observe an absolutely isotropic radiation. However, the degree of anisotropy one anticipates is rather small—in the neighborhood of one part per thousand. After the initial discovery of the microwave radiation, efforts were made to detect this anisotropy. Indeed, an anisotropy of the order of 1 part in 1,000 has been measured. From this measurement we can conclude that our local group of galaxies has a net velocity of 550 km/s towards a certain direction with respect to the center of gravity of our expanding universe.

2. After this velocity (of the local group), the motions due to earth's rotation, earth's orbit around the sun, and the sun's motion in our galaxy are taken care of, the radiation is extremely isotropic (better than 1 part in 10,000).

3. Radiation is a form of energy; the equivalent mass density of this radiation is 0.0003 h/cbm, a rather small addition to the density of matter due to galaxies.

Invisible Forms of Matter—Dark Matter

Although an open universe, one with a beginning but without an end, would appeal to many, for a theorist working in the field of cosmology such a universe presents an enormous problem; to cite one, many quantities become infinite upon calculation, a premise not readily acceptable to some theorists. (A closed universe, one with a beginning and an end, would have none of these problems.) As mentioned earlier, the amount of matter required to close the universe is many times greater than the amount of matter present in the form of galaxies. One natural hypothesis is that some matter in invisible form is present and provides the missing mass. Besides, there are other compelling reasons to believe that some forms of invisible matter exist:

1. Stars revolve too fast in some galaxies. A galaxy is a conglomerate of stars, producing a gravitational field that in turn controls stellar motions, just like planets around the sun. From the brightness distribution of the stars in a galaxy, astronomers can deduce how fast stars must rotate around the galactic center. The theoretical galactic rotational velocities, however, are much smaller than the observed values. To explain this anomaly, one has either to assume that the stars in these galaxies are underluminous (more massive than stars of our galaxy for the same amount of brightness), or that there are some invisible forms of matter contributing to the mass of the galaxies.

2. The same argument may be applied to galactic clusters. A galactic cluster is a conglomerate of galaxies bound by the combined gravitational field of the galaxies. Again the predicted differential velocities are smaller than those inferred from the masses of the galaxies—which are, in turn, inferred from the brightness of the component galaxies.

3. In addition, as discussed later, at the beginning of the big bang process the temperature must have been very high, with particles and their antiparticles coexisting. Later, after the universe expanded and cooled down, these particle-antiparticle pairs annihilated. However, a certain fraction of particle-antiparticle pairs may be left, depending on the rate of expansion (and cooling down) of the universe. If the particles are weakly interacting particles, such as neutrinos, a substantial number of particle-antiparticle pairs may survive the annihilation process. Theory predicts that virtually all neutrino pairs survive the annihilation process.

At present the amount of invisible matter is unknown. Suffice it to say, the possibility remains that invisible dark matter may dominate over the visible forms of matter. We will discuss this later with regard to the evolution of the universe.

General Relativity

We have mentioned general relativity a number of times. It is time now to give a fuller description.

General relativity theory describes the dynamics of particles in a gravitational field, just like the Newtonian theory of gravitation, but in a unique way. In the case of Newtonian mechanics, a mass gives rise to a gravitational field that controls the dynamics of other particles and masses; in the case of general relativity a gravitating mass alters the geometry around the mass, thereby resulting in dynamic motions. To visualize this, imagine a two-dimensional plane made of elastic material, such as rubber. The presence of a mass will cause the rubber sheet to deform into a curved surface. Small test particles will "fall" towards the central deformation and if the deformation is just right, the test particles will follow exactly the same type of motion predicted by Newton's theory. This principle has been successfully utilized to build models of gravitation in science museums. What Einstein did was to prove that the geometrical description of gravitational forces is valid everywhere.

It has been known for some time that Newtonian mechanics is not applicable when the velocity is close to the velocity of light (as Laplace showed in the late eighteenth century); general relativity, on account of the unique properties of the geometry created, can describe dynamical properties even under relativistic velocities (i.e., close to the velocity of light). The geometry (Riemannian geometry) used by general relativity not only includes time as one of its coordinates, it is also different from normal (three-dimensional) Euclidean geometry.

In order that the concept of a Newtonian gravitational potential be replaced by a space-time geometry, all particles must behave in exactly the same way in a gravitational field without regard to their composition. Because geometry makes no distinction between one type of particle and another, all trajectories must be the same. This principle is called the principle of equivalence.

Although Newton did recognize the equivalence of all masses in a gravitational field, the first modern conscientious effort to establish the principle of equivalence was that of R. Eotvos, of Hungary, in 1909. The equivalence principle has since been established to an accuracy of 1 part in 1,000 billion.

The second principle upon which general relativity rests is the principle of covariance. It states that physical laws (specifically the law of gravitation) must be written in a form that is valid in any geometrical configuration (of space and time). To put this in simple language, the physical laws must be like a "universal currency" that can be used anywhere. This requirement appears reasonable, since we are talking about replacing a gravitational field by a geometrical configuration of space-time. Once stated in that way, mathematicians of the last century (Georg Friedrich Bernhard Riemann, 1826–1866, who invented the geometry, and others who worked on it) already had made precise prescriptions of how the geometrical laws—and hence laws of gravitation—should be written.

As we mentioned earlier, the Newtonian laws of gravitation remain valid as long as the velocity is small compared with the velocity of light. Indeed, Einstein's theory of general relativity becomes Newton's theory within the limit of small velocities. Nevertheless there are small differences that are detectable and these differences have been measured with great precision, thus confirming the Einstein theory.

At the time of publication of Einstein's theory of general relativity in 1915, he proposed three tests of general relativity, as follows:

1. He predicted a red shift of light from a gravitating object—the light emitted from the surface of an object, such as a star or even the earth, will suffer a gravitational red shift when observed at large distances from the emitting object.

Because of difficulties in this experiment (one has to be able to measure a redshift in the amount of 1 part in 10,000 from the surface of the sun and 1 part in tens of millions from the surface of the earth) this experiment was successfully carried out only in the 1960s. Gravitational red shift from the sun and from the earth were measured in the same decade, thus confirming Einstein's predictions.

2. Einstein predicted that light, in skimming close to a gravitating object such as the sun, will be subjected to a small amount of bending (change in direction). The bending is also very small, being 1.75 arc seconds near the surface of the sun. The bending of star light near the sun's is most favorably observed during a solar eclipse. In 1918, Sir Arthur S. Eddington organized a solar eclipse expedition to observe the bending of star light. Results of this expedition (and subsequent ones) fully confirmed this prediction.

3. Einstein also predicted that because of the geometrical distortion of space around a gravitating body, planets will not move in strict elliptical orbits as predicted by Kepler and Newton. The deviation is again extremely small, and is exhibited in the form of a slow precession of the perihelion point in the orbit. The planet Mercury, having the most elliptical orbit and being closest to the sun, will have the largest amount of distortion. It had been known for some time at Einstein's time that there is an unexplained precession of the perihelion of Mercury's orbit around the sun, the deviation being 43 arc seconds per century. Einstein's theory predicts this amount in the most natural way.

All in all, there is no doubt about the correctness of the general relativity theory—however, details of general relativity theory are still subject to uncertainties. These details will at most change quantitative predictions of Einstein's theory, but will not change the foundations of the theory of relativity. Nevertheless, these details may affect a number of cosmological consequences.

Cosmology—Principles and Models

Assumptions

Now we are ready to derive cosmological models from Einstein's equations of general relativity. There are tens of equations in general relativity and it may seem impossible to proceed. However, when two assumptions are imposed, these equations become surprisingly simple to solve. These two principles are homogeneity and isotropy:

1. The universe is isotropic—that is, its properties do not depend on any particular direction, and

2. The universe is homogeneous—that is, its properties do not depend on where we observe.

Originally these two assumptions were imposed so that a solution could be obtained (due to simplification of the equations as discussed above). But now a substantial amount of observational data has accumulated supporting these two assumptions. For example, the isotropy property is established through the background microwave radiation to an accuracy of better than 1 part in 10,000.

Cosmological Models

The cosmological models derived under the two assumptions (isotropy and homogeneity) are essentially the same as those of Friedmann and Lemaître. In essence, the solutions are as follows:

1. The universe started as a singular point (whose density is infinite in the context of general relativity), expanding outwards. The velocity of expansion decreases as expansion proceeds.

2. Depending on the density of matter at a particular phase of expansion, three cosmological fates are predicted:

a) If the matter density is below a certain critical value, the expansion will slow down but will go on forever. This is the case of the open universe.

b) If the matter density is just at the critical value, the expansion will gradually slow down and eventually will stop altogether when the universe is infinitely large. This is a particular case of the open universe.

c) If the matter density is above the critical value, the expansion will reach a maximum some time after creation, after which the expansion will reverse to become a contraction. This contraction will bring the universe back to a singular point again. This is the case of the closed universe.

The Cosmological Constant

The three models described above more or less exhaust possible solutions of the Einstein equations. However, related to this solution is an uncertainty that still is not resolved today. This is the cosmological constant.

The Einstein equations are differential equations, which admit a number of constants called integration constants. The cosmological constant is one such constant.

Integration constants in the cosmological solutions are fixed by parameters such as the density of matter, the Hubble constant, and the deceleration parameter (which describes the slowing down of the expansion process at large distances when the velocity of expansion approaches the velocity of light). Indeed, when these three parameters (which can be obtained from observation) are fixed, the cosmological constant is also fixed. However, the current state of observation does not lead to a precise determination of the cosmological constant.

The presence of the cosmological constant changes the quantitative aspects of the three cosmological models, but not the qualitative features. For example, if the cosmological model is a closed one, it will remain closed in the presence of a nonzero cosmological constant. However, the presence of a cosmological constant will change the age of the universe derived from a simple application of Hubble's law and can cause a cosmological model to become temporarily static, although this static state is unstable. (Indeed, this was the motivation for Einstein to introduce a nonzero cosmological constant to reconcile with the then accepted belief of a static universe, an act which Einstein later called a blunder.) In a rigorous model the cosmological constant should be included. However, the uncertainty due to the presence of a cosmological constant is far less than that due to the density of matter, Hubble's constant, and the deceleration parameter. As a result, the cosmological constant has been systematically ignored in most models to avoid unnecessary complications.

The Singularity at the Moment of Creation

Earlier we mentioned that according to Einstein's theory, the universe started from a mathematical singularity. This presents a serious dilemma. There are many indications that Einstein's relativity theory must be replaced by some other nondivergent theory at the singularity. One such effort is the inflationary theory. Although the theory does not go back to the moment of creation, it provides us with a glimpse of what might happen at the moment of creation (singularity). This will be discussed later.

As mentioned previously, enormous theoretical difficulties exist regarding the case of the open universe. However, the density of observed matter in the form of visible galaxies is far below that needed for a closed universe. We will discuss below the role played by invisible forms of matter.

Physical Processes During Creation

These physical processes constitute the most exciting topic of current interest. In 1949 Gamow and Alpher studied cosmological models from the physicists' point of view. Starting from the singularity of creation, they worked out the physical processes that followed. At first there was the very high temperature state during which particle-antiparticle pairs coexisted. Later, as the universe expanded and temperature dropped, the particle-antiparticle pairs began to annihilate, although the annihilation process can never be complete. The number of particle pairs left behind depends on the time scale of expansion and the strength of interaction. Among particle pairs created are proton-antiproton pairs, neutron-antineutron pairs, electron pairs, neutrino pairs, and so forth. While the majority of neutrino pairs survived, most other particle pairs were almost completely annihilated. Conceivably some other not yet detected weakly interacting particle pairs also survived.

As the temperature cooled down to approximately 1 billion degrees, protons and neutrons could interact producing deuterium, a hydrogen isotope. (Deuterium can further interact to produce helium.) According to a number of calculations, during this stage—lasting only a few minutes—from 20 to 25 percent of matter was converted into helium, and virtually nothing heavier than helium. This helium composition would be present today even if no heavier elements were formed inside stars. Even the oldest stars should show this amount of helium in their composition, assuming the correctness of the big bang cosmology theory. Unfortunately helium is one of the most difficult elements to detect in stars. So far, however, no observational data contradict this conclusion.

As the universe cooled down further, nothing drastic happened. With the temperature of the universe falling below 10,000° K, matter previously in the form of free electrons and protons (and helium nuclei), combined into neutral matter. Below this temperature, radiation and matter no longer interact. As the universe expanded further, radiation also cooled down. Eventually this radiation cooled down to 3° K, the background microwave radiation observed by Penzias and Wilson in 1964.

Galaxies probably had not been formed yet and matter in the universe—hydrogen and 20 to 25 percent helium—still was in the form of a homogeneously distributed gas. Here there is some dispute on the state of events: in one theory, when the temperature of the universe was in the neighborhood of 100° K, conditions became ripe for subcondensation of gas into lumps that eventually formed the galaxies we observe today. Others believe that condensation into galaxies occurred much earlier—even before the recombination of gas at 10,000° K. We do not really know what happened.

As galaxies evolved further, more subcondensations took place, in the form of star clusters or individual stars, and eventually the galaxies became what we observe today.

Unresolved Problems

Although we have come a long way from the early theories of the universe, there are still a number of perplexing and unanswered questions, some of which will be discussed below.

Homogeneity and Isotropy

As we have seen, observational evidence points to the fact that the universe is extremely homogeneous and isotropic. How was this state of homogeneity and isotropy attained? If we mix a number of different ingredients in a vessel, it takes great effort to obtain a homogeneous mixture. In particular, we have to bring ingredients from one part of the vessel to the other—back and forth—to produce an even mixture. Likewise, if the universe is so isotropic and so homogeneous, a generous mixing must have taken place in the past to produce this homogeneity. However, nothing can travel faster than the speed of light; and in cosmological models we have today, there was never enough time for light to travel from one end of the universe to the other. How then did this mixing take place?

A new theory, called inflationary theory, originated in 1981 by Alan Guth, deals with this and other unsolved problems. According to this theory, the mixing must have taken place well before the universe was even bigger than a punctuation mark. As the universe evolved from the moment of creation, around a ten billion billion billion billionth of a second later, the universe went through a phase transition during which a rapid expansion took place. Mixing took place at this time, producing the observed homogeneity and isotropy. Subsequently the universe cooled down and expanded to the present state.

Today the inflationary model has only the shape of a skeleton structure. It may not even be the right theory. However, it is the first comprehensive attempt to correlate particle physics with cosmology, to explain the origin of the properties of isotropy and homogeneity. It is based on a particle physics theory, called the Grand Unified Theory (GUT). GUT makes definite predictions, such as a finite lifetime of the proton, something that is now being experimentally studied.

What happened to the universe just before the critical ten billion billion billion billionth of a second? The density was so high, and the particle energy was so great that none of present theories apply. However, it appears from our present knowledge that the four forms of interactions we now know (electromagnetic, weak, gravitational, and strong) were one, a unified interaction for which Einstein was searching during the later part of his life—an interaction that we have only a weak glimpse of at present. Is the Grand Unified Theory the one we are after? Is this interaction beyond our ability of comprehension? Have we reached the Force of creation of our universe? These are certainly questions of the next decade, and probably of decades to come.

Closure Properties of the Universe

We mentioned previously that a closed universe is needed to eliminate mathematical difficulties of the theory. Another compelling reason is that the inflationary model is applicable only to a closed universe. Observationally, we see many inconsistencies if we take the masses of the galaxies at their face value (as obtained from the brightness distribution of stars within a galaxy). Among these inconsistencies are: the stars in galaxies rotate too fast around the galactic center, the mass of galactic clusters appears to be too small to bind the galaxies together, and so on. To remove these inconsistencies, the existence of some forms of invisible, dark matter is postulated. The required dark matter will probably not be ordinary matter. In all likelihood the dark matter will consist of nearly noninteracting particles (like neutrinos), a number of which have been postulated but without experimental confirmation. In addition, there is an inconsistency in the theory of formation of galaxies. If we apply conventional theories to explain the formation of galaxies, we obtain too small a mass. On the other hand, if dark matter exists, galactic formation can take place much earlier in the evolutionary scale of the universe, and with much greater masses.

The next question is, What forms of dark matter and how much? The question "how much" is easier to answer. If we want to close the universe, an equivalent of at least 3 h/cbm is needed. When compared with the density of matter due to galaxies, which is at most 0.2 h/cbm, at least an additional 2.8 h/cbm equivalent of dark matter is needed. The existence of this amount of dark matter—at least 14 times that of the ordinary matter that constitutes the galaxies, stars, earth, and last but no the least, man, further diminishes man's position in the universe. If dark matter exists as postulated, man has to accept the fact that he is an insignificant part of the universe.

There are only speculations as to what forms of dark matter should exist in the universe. One possibility is that neutrinos comprise 90 or more percent of the dark matter in the universe—there is some experimental evidence that the rest mass of the neutrino is not zero. Other explanations that concern hypothetical particles, such as photinos and axions, are in theory equally valid.

Limitation of Our Abilities

So far I have given the impression that our reasoning power appears to be unlimited and the contents of our theoretical toolbox can increase without limit. In reality, neither is true. Our reasoning power relies upon our ability to gather facts relevant to the problem. We have only limited vision to probe the universe, and there are secrets of the universe that are apparently beyond our meager physical means, at least for the present time.

Firstly, our ability to probe the universe is theoretically limited by the so-called horizon. On account of the curvature of the earth, even the tallest tree or building will eventually vanish below the horizon and become invisible. According to Einstein's theory of relativity, such a horizon also exists in the universe—called the particle horizon or event horizon, which is the maximum distance a photon could have traveled from the beginning of the universe to the present. Objects close to our horizon will appear to us to have suffered nearly infinite redshift, thus reducing their apparent brightness to zero.

Secondly, long before the event horizon places a limit on our vision, our views are blocked by another mechanism. Virtually all our knowledge regarding the universe is derived from observations in the electromagnetic spectrum—from radio waves, to microwaves, to infrared, to visible, to x-rays, and gamma-rays. During the evolution of the universe, there was a stage when the protons and electrons recombined into neutral hydrogen atoms—an event that took place when the temperature of the universe was 10,000° K. During this recombination period, photons were absorbed and re-emitted many times. Any electromagnetic information emitted prior to this epoch will have been completely scrambled. If we could look back to this time, the universe would appear to us as a homogeneously illuminated entity, as indeed it does. The 3° K microwave background radiation we detect is both homogeneous and isotropic. This is the radiation left behind in the universe after hydrogen recombination was completed, when the universe was only tens of thousands of years old. *No matter how hard we try, we cannot see objects younger than this epoch.* This means we can only probe the evolution of our universe prior to this epoch by reasoning, given whatever scientific facts we can find.

Through reasoning, using tools such as nuclear physics, thermodynamics, and others, we concluded that helium was synthesized in abundance (25 percent) during the first few minutes of creation. We have fairly strong evidences for this conclusion.

Prior to helium formation, we have to rely heavily on our knowledge of particle physics, which has taken enormous strides during the past several decades. Particle accelerators continue to grow in size to the point that the cost of a recently proposed accelerator becomes a visible fraction of the national budget of this country. Yet we are far away from what is required to test the inflationary model, the physics at the first ten billion billion billion billionth of a second. The properties of the inflationary models are based on our knowledge derived from what may appropriately be called the low energy behavior of particle physics, even at the highest energy attainable on the earth. Alan Guth, the proponent of the inflationary model, estimated that in order to test this theory at the particle physics level, two accelerators, each several light years long (using present-day technology), would be needed. Such tools are forever beyond our ability, unless we learn to manipulate masses the size of the stars with ease, as we do with the cranes, bulldozers, transistors, and integrated circuits of today.

Have we reached the end of our search? In the view of our current technology, we apparently will do so soon. Nevertheless, scientific and technological progress of the past several decades have taught us a lesson: he who prognosticates future upon present shall be punished with ignorance. Solutions do not go to those who stop searching.

Conclusion

In this paper I have reviewed aspects of cosmology, which have been under development for the past several centuries. Our current concept of the universe is a far cry from the primitive view held in medieval times, that is, that the earth comprises the entire universe. The age of the universe was expanded from the biblical value, around 6,000 years, to approximately 20 billion years, according to the most modern view. Not only have we been able to reconstruct the events that occurred almost at the time of creation, we have also been able to prove our reconstructions by observations and experiments.

At present it appears that we have almost reached the moment of creation. However, as we take one further step towards the moment of creation, our step size seems to have diminished by almost the amount we have advanced. We have now reached perhaps the first second of creation, and if the inflationary theory is successful, the first ten billion billion billion billionth of a second, but we no longer count time by seconds but rather in units so small that they are beyond our ability to measure. Yet time has meaning in the universe itself; the universe has to cross that seemingly infinitesimal time interval in order to evolve to the one-second state. The evolution of the universe is its own clock. Are we to cross the seemingly infinitesimal time gap to reach the moment of creation? Can we ever cross this final barrier? Probably not. However, man is a curious being. As long as there is the least amount of the unknown left to be searched, he will continue his search.

Ten years ago cosmologists talked about the state of events at the first minute of creation. Now we have reached the threshold of ten billion billion billion billionth of a second, although we have yet to cross it. Ten years from now we may still be trying to cross this threshold, or we may have crossed it and find ourselves confronted with still another unknown barrier.

No doubt the next decade should be the most exciting one as far as cosmology is concerned. The launching of the Space Telescope—appropriately named the Hubble Telescope—will bring back data of undreamed of quality. The amount of work done in the past decade on theoretical cosmology is probably equivalent to several centuries of work done in the past. Coupled with the forthcoming data from the Hubble Telescope, the next decade should prove to be even more productive. We probably will never reach the moment of creation but we will always continue our efforts to reach it.

In a separate paper of this volume, Dr. Sebastian von Hoerner, a pioneer in radio astronomy and search for extraterrestrial intelligence, discusses three categories of cosmologists, those holding prevailing views (as discussed in this paper), those holding a more conservative view (that we must base all our deductions on observations), and those holding daring, original, and often wrong views. I would classify the first category as the establishment, the second category as the conservatives, and the third as the radicals.

While the past has seen many converts from conservative to establishment, very few reverse conversions took place. Although in this article I have touched on relatively few ideas of the radical group, it is this group that I now salute. It is not because the radicals are right. On the contrary, 99 percent of this group are probably wrong. However, it is from this group that our present views emerged—the ideas of Copernicus, Bruno, Galileo; even Einstein, Lemaître, and Friedmann. Today we have the inflationary proponents. There is a large probability that the present formulation of the inflationary theory may not survive, yet the theory must not and cannot be ignored. From efforts like this we shall acquire the cosmology of tomorrow. This is not the only radical group worthy of mention. There are others, like those holding different views, especially on the nature of the redshifts for quasars and certain types of double galaxies.

Cosmology differs from other branches of science, and even other branches of astronomy, in that very few events can be linked to the moment of creation. To utilize these few events, enormous tasks must be performed, using many disciplines of science. The final conclusion is still uncertain, and depends on the availability of evidence and human judgment. I would like to compare cosmology to the work of anthropologists attempting to trace the beginning of the human race. Fossils are few and far between, and to study each piece of evidence, the most modern and advanced technologies in every discipline useful to their work have been applied. Yet at the end, it is not the machines nor the technologies that generate the conclusion, it is the human mind that passes the final verdict. Human verdicts may be wrong. Yet every conclusion brought forward advances us to another new dimension in our view of the origin of our species. An anthropologist cannot but be in awe at the few fragments of bones from which great conclusions are drawn and also from which he or she might possibly have emerged. A cosmologist, when looking at a piece of evidence, which is much more abstract than the counterpart of the anthropologist, has the same awe: could this photon he just detected be the one that the atoms of his body interacted with at the very beginning? He will never be able to verify this fact, yet each time he looks up at the sky, at the stars, he is rewarded by the sheer beauty of the heavens that only a human can relate—somehow, the universe did not forget to give him the ability to be human.

Acknowledgments

I wish to thank Dr. Fred Singer for the extensive amount of time he spent in reading this manuscript and subsequent discussions.

REFERENCES

Cosmology is such a vast subject that it is hard to give an adequate reference list. Many original works dealing with early cosmological models were written in obscure style with heavy emphasis on mathematics. The following references give comprehensive coverage of topics covered in this article.

Gibbons, G., S. Hawking, and S. Siklos. *The Very Early Universe.* Cambridge: Cambridge University Press, 1982. The authors are themselves forefront workers in cosmology. Hawking, for example, is noted for his discovery of a decay mechanism for black holes via a radiation process that bears his name. This volume is the proceedings of the Nuffield Workshop in Cosmology that was held in Cambridge, June 21 to July 9, 1982, and is not for laymen. It contains detailed descriptions of all modern concepts of cosmology, including the inflationary universe. A more recent, and quite popular book by Hawking is *A Brief History of Time.* New York: Bantam Books, 1988.

Munitz, Milton K., ed. *Theories of the Universe.* New York: Free Press of Glencoe, Macmillan, 1957. This book contains original essays ranging from ancient times to the present year, with many personal notes by the cosmologists involved. Authors include: Ptolemy, Copernicus, Bruno, Galilei, Kepler, Newton, Einstein, Lemaitre, Gamow, Bondi, Hoyle, and Eddington.

Weinberg, S. *The First Three Minutes.* New York: Basic Books, 1977. A well-known particle physicist describes the state of events from the first one-hundredth second to the first three minutes of our universe. Element synthesis, fireball (leading eventually to the 3° K background microwave radiation), as well as other topics, are discussed. Weinberg's work is strongly related to 1980s cosmology, which includes the theory of the inflationary universe.

11

Comment on Hong-Yee Chiu

SEBASTIAN VON HOERNER

Hong-Yee Chiu's paper presents a good general background, and puts the emphasis on modern theory and its results. My discussion paper adds some comments, changes some perspectives, and puts the emphasis on difficulties, problems, and oddities.

Among others the following examples are discussed: our difficulty of disentangling the thorough mixture of space, time, and objects as received in all our observations; very unlikely but observed close pairs with different redshifts; the matter-antimatter problem; the infinite starting velocity of the big bang, which causes the so-called "mixmaster problem"; our inability to tell the future after the final collapse; and the similarity of relativistic and Newtonian models, in spite of their different (mutually exclusive) foundations. It is most amazing (and unexplained) that the universe is not too different from the simplest of all possible universes, the Einstein-de Sitter model.

Hong-Yee Chiu's paper about modern cosmology is a very good summary of the prevailing picture of the universe, as shared by most of the present workers in this fascinating field. It includes some background about history and observations, and describes well the importance of theories, our "box of tools of reasoning." The paper needs only few critical remarks; thus, I want mainly to make additional comments and to show different perspectives.

Regarding cosmology, we have three groups of scientists. First, the daring spirits who ask what kind of universe is possible at all, and who calculate various model universes. Second, the observers, who want to stay closer to what can actually be seen, and try hard to push observational limits ever farther out. Third, some sceptics choose to regard the results of the first group as mere speculations, and those of the second as misinterpretations arising from observing errors. All three attitudes and their interactions are needed in our search for the truth. Since I know Chiu as a good, active theoretician of the first group, I will put my emphasis on the difficult, the unknown, and the odd items of cosmology.

Cosmology faces many difficulties. Three have already been well described by Chiu: the general horizon of the universe beyond which nothing can be observed; the lack of any signals from the time before the emission of the 3° K background radiation, which means that all events of the first 100,000 years of our universe can never be observed directly, but must be deduced by reasoning; which leads to such extremely high temperatures that we must assume new states of matter that can never be tested experimentally.

Three more difficulties need to be mentioned. First, size and quality of telescopes and equipment have technical or financial limits, and we face either severe degradations by looking through our atmosphere or extreme expense for space telescopes. Second, we observe our objects against unavoidable cosmic background noise: radiation from gas clouds of our own galaxy, numerous faint faraway galaxies, and the so-called photon noise (radiation is not a continuous flow but consists of single photons with finite energy). Third, we want to ask single well-defined questions, but the observational answers are always an odd threefold mixture: looking far out into *space* (asking for its curvature for example) we necessarily look far back in *time* (when all was different); and we do not see the universe itself, but only *objects* (e.g. stars and galaxies, which had an evolution of their own). Disentangling this odd mixture is one of our main problems.

I believe that the most basic difficulty, but also a great challenge, is the fact that cosmology, like all astronomy (except planetary radar), is a "passive" science, since we see only whatever signals happen to come our way, and are unable to force "active" experiments. Hong-Yee Chiu compares cosmology with anthropology in their search for the origin. I would like to stress this similarity by quoting Alan Walker who said that anthropology is "like a three-dimensional jigsaw puzzle with no picture on the box and half the pieces missing." Our most basic limitation is a more philosophical one, illustrated by the question "What was before the big bang?" On this matter, I quote Martin Luther; when asked "What did the Lord do before the creation?" he answered, "The Lord was sitting in the bush, cutting sticks to beat those people who ask such silly questions!"

Because of these and other difficulties, we actually know rather little about the universe. We know it expands, but the rate of expansion versus distance, the Hubble parameter, is not just 75 km/sec per megaparsec as often quoted; our best observers still disagree as to whether it is only 50 or as high as 120. This means that cosmic distances are uncertain by a factor of two, which means the cosmic density of the observed galaxies is uncertain by a factor of $2^3 = 8$. But the average galactic mass is also uncertain by about a factor of four (curves of rotation velocity indicate a lot of thinly distributed mass in the outer parts), making such a simple quantity as the observed average density uncertain by at least a factor of 30. And on top of this we have the indication for, and uncertainty of, the dark "missing mass" as discussed well by Chiu.

It gets even worse if we go one step further, asking for the change of the expansion, that is, the deceleration parameter q. Present observations of redshift versus distance can limit it at best to between zero and one, a large range of uncertainty, which is larger still for the mostly omitted cosmological constant Λ, as explained by Chiu. I should add that the simplest of all world models is the Einstein-de Sitter model, with $\Lambda = 0$ and $q = \frac{1}{2}$, and with just the critical density for parabolic expansion (decelerated to rest after infinite time) within a flat (uncurved) space. Because of its beautiful simplicity, this model is frequently used, or even assumed to be true.

This calls for a comment. Our preference for simplicity and beauty may be a good guideline (in lieu of more solid arguments), and frequently led to good results; but any conclusions, before a convincing test, must be regarded with caution. Also, a lot depends on personal opinion. Hong-Yee Chiu would like to reject any open universe, because of its infinities (size, mass, energy . . .), while I feel much troubled by the infinities at the start of the big-bang (density, temperature, expansion velocity). But we should remember that the ancient Greeks found nothing but circles acceptable for planetary movements; that the steady-state theory was developed for its beauty; and both were later contradicted by improved observations.

One thing we know is that at early times radio galaxies were more numerous, more luminous, or both, and that there were also more quasars. These object evolutions make the disentangling of the universe from its objects very difficult indeed; but they also yield a strong argument against the steady-state theory, together with that from the 3° K temperature background radiation mentioned by Chiu.

In addition to uncertain observations, we also have some rather uncomfortable observations that deserve to be at least discussed before, as usual, being pushed under the rug. Halton "Chip" Arp of the Mt. Wilson Observatory has collected a number of close pairs of quasars and galaxies (both mixed and one-type pairs) that have very different large redshifts. Since large redshifts are always taken as a measure of large distance, via the cosmic Hubble expansion, the two members of such a pair then would have very different distances from us along the line of sight; thus the observed closeness of the members would only be an apparent one, a chance effect after their projection on the sky. But Arp claims, and his data seem to support it, that the closeness of these pairs and their numbers go far beyond any reasonable chance expectation. This would mean that redshifts are not, or are not always, a measure of distance, or even of velocity. Quasars then could be nearby and less luminous, but all detailed models and explanations so far have failed. I must leave this an open question.

We have more oddities to deal with. In all our terrestrial experiments and observations of cosmic rays, whenever matter is created, it is always created absolutely symmetrically, with matter and antimatter in equal numbers—each particle simultaneously with its antiparticle. Since we always extrapolate from the lab to the universe, mostly with good success, and since symmetries play a decisive role in modern physics, we should assume the same symmetry for the big bang—simultaneous creation of particles and their antiparticles. But then we have to explain: why are our earth and Solar System, certainly our Galaxy and at least all the galaxies of our local group, made up of ordinary matter only? Explanations using a separation of matter and antimatter in large blobs at an early time have been tried but look very unsatisfactory. But so does a slight asymmetry at the big bang. Again an open question.

The problem of the origin of the observed homogeneity and isotropy has been discussed by Chiu: "If we mix a number of different ingredients in a vessel, it takes great effort to obtain a homogeneous mixture." Since big bang models have a finite horizon, becoming smaller at earlier times and even going to zero at time zero, there was no time and opportunity for effective mixing. Chiu then mentions the new inflationary theory by Alan Guth, which allows enough mixing during a very early and fast-expanding phase of the universe. This, as well as some earlier treatments of this so-called "mixmaster problem," tries, in my opinion, only to remove the symptoms of a much more basic illness: the zero-horizon at start, which means there was *no* causal connection between any two parts of the new-born universe, thus leading to the concept of different ingredients at different places that require mixing. The real problem is not the mixing; it is this "common but unrelated origin of all things," as I have once called it, completely unacceptable to my intuition. It results from the zero-horizon, which results from starting the big bang with infinite velocity. We should look for a model without such extreme starting velocity. Then a common origin can be causally connected (as it well should be) and may produce the *same ingredients* everywhere in a natural way.

It is also odd that our physics is unable to tell in principle what comes after the final collapse of a closed universe—just as it cannot tell what happens inside a black hole after its gravitational collapse. According to our physics, neither the universe nor a black hole can "bounce" up again. There cannot be an "oscillating" universe with a series of big bangs, expansions, and collapses. But we do not have any alternative either!

One more oddity. Before Einstein, a classical or Newtonian cosmology was developed that considered the negative potential energy of mutual gravitation and the positive kinetic energy of the universal expansion inside some large arbitrary volume of the universe. It was found, depending on the total energy being either positive, zero, or negative, that there are three types of universe: expanding forever, coming to rest after infinite time, or reaching a maximum expansion and then collapsing again. The odd thing is that the results of Newtonian and relativistic models are strikingly similar (if $\Lambda = 0$), whereas the basic considerations are completely different. Einstein considers only the rest-mass energy ($E = mc^2$) and the thermal energy (pressure), both omitted in Newtonian cosmology; Einstein omits both the potential energy and the kinetic energy of expansion, as can be seen from his field equations and the energy-momentum tensor. Let me try a simple explanation. If we compare energies in an arbitrary volume, say a sphere of radius r, then this seems possible only if all those energies vary with the same power of r. Now, the Newtonian potential and expansion energies both vary with r^5, whereas the relativistic rest-mass and pressure energies both go with r^3 (and can be used also locally as done by Einstein). Thus, one may do it either way, but should have a bad conscience about the terms omitted by Einstein. No omission is needed only if one considers the total energies of a closed finite universe.

Finally, a nice non-trivial conclusion follows from our ignorance. We cannot tell the sign of space curvature and cosmological constant. Thus both cannot be too different from zero, and our universe cannot be too different from the simplest one, the Einstein-de Sitter model, the only one where Einstein *and* Newton give the same results.

REFERENCES

Good references about modern cosmology are already provided by Hong-Yee Chiu. Regarding the comparison with the older Newtonian cosmology, I would like to add the following:

Bondi, H. *Cosmology.* London: Cambridge University Press, 1960. This book also gives a good discussion of the basic philosophies of various theories of cosmology.
McVittie, G. *General Relativity and Cosmology.* Urbana: University of Illinois Press, 1965. This book provides a wealth of useful equations, with short, understandable derivations.
Walker, A. *National Geographic Magazine.* vol. 168, no. 5 (1985): 610. This article was the source of my quotation about anthropology.

THE ORIGIN
OF THE
SOLAR SYSTEM

12

The Origin of the Solar System

JOHN S. LEWIS

The Solar System comprises many different classes of bodies with astonishingly different chemical and physical properties. By what processes did these bodies arise from a largely homogeneous initial state? This review summarizes our theoretical attempts to solve this problem, and presents a brief and selective review of recent data on our planetary system as a means of testing the predictions of these theoretical constructs. Some crucial remaining uncertainties that hinder us in choosing between these theories are mentioned, and the prospects for early testing of these theories are discussed in light of present plans for planetary exploration in the United States, the Soviet Union, Europe, and Japan.

Astronomical Setting of Star Formation

A large body of astronomical observations suggests that the formation of stars and stellar systems begins in the gas- and dust-rich lanes in the spiral arms of our galaxy. There the interstellar gas and dust clouds can attain densities so high that their gravitational potential energies exceed their internal thermal energies. When an interstellar cloud (compressed by an intercloud collision, a passing shock wave from a stellar explosion, or the radiation pressure of surrounding stars) becomes sufficiently dense to meet this criterion, its internal pressure ceases to be sufficient to prevent its collapse.

Collapsing interstellar clouds may commonly be formed with a large amount of angular momentum. Their collapse can then cause spin-up to such high speeds that they fragment into dense cloudlets orbiting about a common center of mass. The original cloud may have a mass of several thousand suns. After several generations of collapse and fragmentation, and after a time measured in hundreds of thousands of years, a hierarchy of structures will be found, ranging from individual prestellar nebulae with masses roughly comparable to that of the sun, through small associations of only a few such nebulae, on up to clusters of hundreds to thousands of nebulae. Indeed, the spiral arms of our galaxy today are full of such associations and clusters. The large majority of the stars in the spiral arms (outside the dense, gas- and dust-poor galactic core) are indeed found in association with other stars. Only 5 to 20 percent of the stars near the sun are apparently single stars, devoid of stellar companions. The rest belong to double, triple, and even more complex multiple stellar systems. Since some stars are exceedingly faint, with luminosities ten thousand times smaller than that of the sun, many of the stars that appear single to us may in fact be accompanied by one or more very faint companions.

The somewhat unusual status of the sun as a single star does not necessarily mean that planetary systems like our own are infrequent. Many stellar systems are either very compact, with the stars much closer together than Mercury and the sun, or very widely separated, with orbital periods of thousands of years or more. In such multiple systems there is a wide range of separation distances within which planetary orbits would be stable for billions of years. In many systems, two or three stars orbit so far apart that each star could have its own stable planetary system. The masses of the stars in multiple systems range from a few percent of the mass of the sun (so-called "brown dwarfs," the smallest bodies with enough luminosity to be visible to us over interstellar distances) up to an upper limit that lies somewhere in the range from 60 to 200 solar masses. The luminosities of the stars vary as a very sensitive function of their mass, from about 0.0001 solar luminosity for the least massive to roughly 1,000,000 solar luminosities for the most massive. The sun is brighter (more luminous) than about 95 percent of the stars. The most common type of star in the galaxy is the red dwarf, a tiny, cool stellar type with less than a thousandth of the sun's luminosity. The overwhelming majority of the mass in the spiral arms of the galaxy is in the form of dark or faintly luminous matter, such as red dwarfs and interstellar matter. On the other hand, the minuscule portion of the total stellar mass that is found in very massive stars is responsible for virtually all of the luminosity of the galaxy.

Two stable, hydrogen-fusing stars that differ by a factor of 100 in mass (say, 0.1 and 10 solar masses) will differ by a factor of several million in luminosity. Thus 10 solar masses of material distributed among 100 equal-mass small stars will contribute about 100,000 times less light than it would if collected in a single star. It requires little mathematics to deduce that the hydrogen fuel in the larger stars will be consumed in a very short time, whereas low-mass dwarf stars can survive for hundreds of billions of years, a time far longer than the 15-billion-year present age of the universe.

When low-mass stars exhaust their supply of fuel (in the distant future), they will subside gently into obscurity. But very massive stars have such powerful gravity that they can build up and hold immensely high temperatures and pressures in their interiors. In the deep interiors of low-mass stars, conditions are barely suitable for the slow fusion of hydrogen into helium. But at the extreme temperatures, pressures, and densities encountered in the deep interiors of massive stars, a host of complex nuclear reactions can take place, leading to the synthesis of a wide range of chemical elements ranging from helium through carbon, nitrogen, and oxygen, then the rock-forming elements silicon, magnesium, and iron. The evolution of such a massive star constantly accelerates until, in a final incredibly violent paroxysm, the star explodes. A portion of the star, containing every known element up through the radioactive actinides, is ejected in a powerful shock wave traveling at a few hundredths of the speed of light. For a brief time, the blazing remnants of the exploding star shine brighter than an entire galaxy. The ejected shell of gas departs, leaving behind an immensely dense sphere of collapsed matter where the star once orbited.

A star with 30 solar masses will burn its fuel about 200,000 times as fast as the sun. Instead of remaining a stable, well-behaved hydrogen-burning ("main sequence") star for 10,000 million years like the sun, it will instead run through its entire lifetime in a mere 50,000 years. This means that in regions of star formation, where stars will appear over a time of hundreds of thousands of years, any very massive stars formed early in the collapse process will have run through their entire lifetimes and exploded while star formation is still taking place nearby. Indeed, the high luminosity of the star pushes against nearby gas and dust clouds and accelerates their collapse. The star's final cataclysmic supernova explosion not only helps squeeze the interstellar cloud material to high densities, but also fecundates it with freshly synthesized elements. The rock-forming elements are the raw material of terrestrial-type planets; the radioactive elements are the fuel that drives the melting, thermal evolution, outgassing, and internal convection of the planets; and the carbon, nitrogen, oxygen, and other light elements are the precursors of life.

Formation of Planetary Systems

Let us now consider a region within which star formation has been occurring for some time. Many stars, mostly tiny red dwarfs, have come into being. A small proportion of the new stars have masses similar to that of the sun. Collapsing prestellar nebulae have shrunk to sizes comparable to the present size of the Solar System by radiating off their internal energy. Angular momentum has been shed by fragmentation episodes and by the shedding of mass from the outer edges of the collapsing nebulae. Each rotating nebula collapses rapidly toward its equatorial plane, but its angular momentum is sufficient so that, during collapse, the equatorial regions of the nebula spin up to high speeds and eventually approach orbital velocity. This centrifugal force very strongly inhibits further collapse toward the spin axis. Each flattened, collapsing disk shrinks so as to flatten itself further, accretes additional matter onto its surface, and radiates heat from its surface into space as its interior heats up.[1]

The nebula takes on a flattened disklike shape, symmetrical about the spin axis. Rarely, a nebula with 20, 40, or even 60 solar masses collapses far enough to generate temperatures high enough for the fusion of hydrogen, and a highly luminous main sequence star is born. This star lights up, blows back the surrounding gases and any dust that has not agglomerated into larger bodies, runs through its entire evolutionary lifetime, and then explodes into the surrounding interstellar clouds. Any prestellar nebulae formed out of that shocked and chemically enriched cloud material will be chemically, elementally, and isotopically complex and heterogeneous.

Astronomical observations show many objects along the sequence from interstellar diffuse clouds to dense, flattened nebulae, but these dense prestellar nebulae are small targets and faint emitters, and tend to form in parts of the sky that are already stunningly complex and full of spectacular, more easily seen features. Moreover, the nebular stage in the evolution of a star is only a tiny fraction of the star's life: as a rule, a sun-like star spends only 0.0002 percent of its life in the nebular stage. For these reasons, few observations exist pertaining to the evolution of a flattened nebular disk into a planetary system.

Theory strongly suggests that within such a flattened disk dust particles must collide with each other so frequently that, even if the probability of sticking is small, the dust particles will rapidly agglomerate into meter-sized dustballs. These bodies, attempting to orbit the central star (hereafter referred to as the sun), find themselves embedded in a gas with a pressure of a millionth to a thousandth of the earth's present atmospheric pressure. Any initial motion that causes them to depart from the equatorial plane of the nebula will rapidly be damped out by gas drag, and the accreting bodies will soon be strongly concentrated in a very thin dust layer in the central symmetry plane of the nebula. There, the gravitational influence of this very dense dust disk will be so powerful that the small bodies will rapidly accumulate into asteroid-sized (100 to 1,000 km) bodies.[2] At some point late in this evolutionary sequence, temperatures and pressures in the center of the nebula become so high that nuclear reactions begin, the central star lights up, and it rapidly disperses into space any gas and unaccreted dust that may be left in the nebula. During this early brief stage of the life of a star it is excessively luminous and emits a powerful stream of protons and electrons a million times stronger than our sun's present solar wind. This stage in the life of a young star is called the "T-Tauri phase" after the first known star of this type.

The further accretion of solid material occurs in accordance with the laws of celestial mechanics. The solid bodies occasionally collide and accrete to form larger bodies. Rarely, they collide sufficiently violently to disrupt each other. Very commonly, they pass close enough to each other so that their gravitational interactions perturb both of them. Such interactions have the effect of pumping up the orbital eccentricities and inclinations of the smaller bodies, which forces them to cross the orbits of a greater number of the larger bodies, which in turn hastens their accretion. Theory tells us that the gas-free accumulation of asteroidal- and lunar-sized bodies into rocky planets takes place over a time scale of about 100 million years.[3,4,5]

The materials available for planetary formation are by no means uniform in composition throughout the nebula. Just before dissipation of the gas and fine dust, temperatures in the deep interior of the solar nebula must have been high enough to vaporize rock completely not too far inside the present orbit of Mercury.[6] A little farther out, only the most refractory (involatile) elements and minerals, such as tungsten, iridium, and minerals dominated by oxides of aluminum, titanium, and calcium would have been condensed and available for accretion into large solid bodies (see Fig 12-1.) Yet farther out, the very abundant elements—namely, iron, magnesium, and silicon— would have joined the condensate, largely as metallic iron-nickel alloy and the magnesium silicate mineral enstatite. Yet farther from the center of the nebula, relatively volatile rock-forming elements such as sodium and potassium, the halogens chlorine and bromine, then sulfur, and finally chemically bound water (such as in clay minerals) will join the condensate and be available for incorporation into newly forming planets. At great distances from the center, temperatures will be so low that ice minerals such as water ice, solid hydrates of carbon dioxide, methane, ammonia, nitrogen and rare gases, and solid carbon dioxide itself will condense and be available for planet formation.[7]

As the temperature drops off with increasing distance from the center, so too do the tidal effects of the gravity of the central mass concentration, which would readily strip atmospheres from bodies orbiting close to the sun. Further, lunar-sized bodies embedded in the inner nebula would be unable to capture the hot gas from the nearby nebula because their gravitational attraction would be small compared to the thermal energy of the gas. At great distances from the center, however, similar-sized bodies would be more numerous (many more elements, and much more of the total mass, is condensed at lower temperatures), and the thermal energy of the surrounding nebular gas, as measured by its temperature, could be up to ten times smaller than near Mercury's orbit.[8] This means that objects little larger than the moon, if present in the outer nebula, could capture vast masses of nebular gas, thus enormously multiplying their original masses, leading to a runaway accretion of gas to form planetary

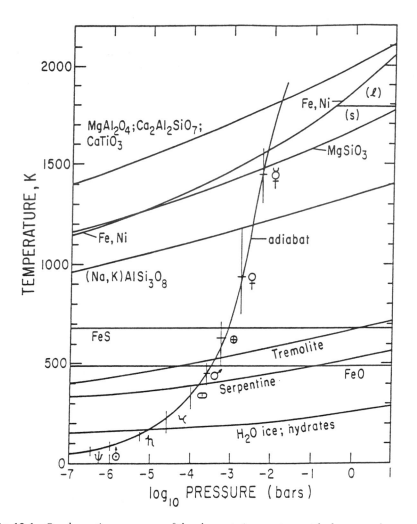

Fig. 12-1—Condensation sequence of the elements in a system with the same elemental abundances as the sun. The diagonal curve is a representative (adiabatic) T-P profile for the solar nebula at the time of the formation of preplanetary solids. The other curves are condensation points of the major components of the terrestrial planets. The regions from which the bodies in the Solar System sample most of their mass are indicated by their symbols: from top to bottom along the nebular adiabat, these are Mercury, Venus, earth, Mars, Ceres, Jupiter, Saturn, Uranus, and Neptune. The high-temperature minerals (those stable closest to the sun) are especially rich in calcium, aluminum, and titanium. Most of the condensate mass is in the form of solid (s) or liquid (l) iron and magnesium silicates, the principal components of planetary cores and mantles. The more volatile alkali metals sodium (Na) and potassium (K) are retained in preplanetary solids only at somewhat lower temperatures, and sulfur is retained only below ~680K. Tremolite and serpentine are water-bearing minerals. The line marked FeO is the endpoint in the progressive oxidation of iron metal: below it, no metal grains survive. Ices are stable only at the very low temperatures found in the outer Solar System. The temperatures are probably the *highest* experienced by the nebular gas dust mixture at each distance from the sun.

118

bodies. These outer planets would therefore not only be tens to hundreds of times more massive that the inner rocky planets, but would also be much less dense, since they would be dominated by hydrogen and helium, not rocks.[9]

This is, in brief, our present qualitative physical and chemical understanding of the way the Solar System came into being. It is an essential part of science that such qualitative descriptions, once found free of obvious internal logical contradictions, be quantified as thoroughly as possible. They must then be subjected to the most exhaustive testing by confronting them with the full range of observed properties of the Solar System as deduced from modern earth-based and spacecraft-based observations. The best of these theories, as judged by their ability to explain observations and make successful predictions of the outcome of future observational tests, will generally be imperfect and require revision. Each successive round of observation, and each new spacecraft encounter with a previously unvisited planet, asteroid, comet, or satellite, exacts its cost from theorists, and rewards them also, sometimes in astonishingly unexpected ways. By this iterative process, the theoretical and experimental feet of science take alternate steps toward the common goal of the understanding of nature.

For our present purposes, the most effective way to provide a more detailed description of present theories is to provide a brief summary of the general properties of the Solar System as presently understood. Such an approach does not do justice to the historical development of our understanding, and leaves unexplained the apparently perverse fascination of many theorists with ideas that now seem to have no explanatory power. Almost without exception, however, even the most transient of these theories was invented for good, if not compelling, reasons to explain some small islands of observations that have now either been incorporated into the mainland, or discredited as being in error for either experimental or interpretive reasons.

Present Properties of the Solar System

We now face the difficult task of summarizing tens of thousands of publications dealing with the origin, evolution, composition, and structure of every class of bodies in the Solar System. Prior to the advent of planetary spacecraft missions and the development of modern spectroscopic, radio, and radar techniques in the 1960s, a very large proportion of our knowledge of the Solar System was derived from the laboratory study of meteorites, the "poor man's space probes." Our understanding of the intrinsic properties of other planets was limited in the extreme: aside from the moon, we had only a few dozen isolated facts at our command to characterize the geochemistry, geophysics, and meteorology of the other planets. Our understanding of the earth, although incomparably better than that of the other planets, did not become truly modern until we attained knowledge of sea floor spreading and global tectonics in the mid 1960s, coincident with the birth of the planetary sciences. Further, even today at least 99 percent of our geological and geochemical data on the earth treat only the outermost, most accessible 0.5 to 1 percent of its mass. This nonrepresentative sampling of the earth leads to its own biases and errors. Our task is to discriminate between those features and systematic trends that are universal to all planets and the individual idiosyncracies of the one planet we happen to know best. It is difficult to the point of impossibility to do so using only detailed data on a small part of one planet and extremely sketchy (and often erroneous) information on the other planets.

Meteorites

The reason that the study of meteorites first came into prominence, and one of the main reasons that research on them continues today, is that they appear to sample a considerable number of parent bodies from a wide range of locations in the Solar System.[10] Many of them have retained their primordial compositions and structures and have ages that show that they were formed at the same time as the planets, some 4,500 million years ago. As such they are the most ancient samples of Solar System material. If we could determine their exact times and places of origin, we could then reconstruct a sort of movie of the evolution of chemical and physical conditions in the solar nebula and in the post-nebular planetary accretion phase. Determining the ages of meteorites is now routine, but reconstructing their places of origin in the Solar System has proved to be an immensely difficult chore.

In the 1960s it was generally accepted that the primitive, ancient meteorite types, the chondrites, were mostly from the asteroid belt.[11] The most common class, the volatile-poor and relatively unoxidized "ordinary" chondrites, were assumed to be from the belt, and the highly oxidized and volatile-rich "carbonaceous" chondrites were regarded as possible samples of cometary material. Estimates of the formation temperatures of these meteorite classes can be made from their measured content of a number of moderately volatile elements. When these temperatures were assigned to the asteroid belt on the assumption that the ordinary chondrites formed there, a very large temperature discontinuity was found between the belt and Jupiter. Further, the formation temperatures of the ordinary chondrites were so similar to that deduced for the earth that one was forced to conclude that the entire inner Solar System was formed at nearly the same temperature. The terrestrial planets would then be formed of virtually identical material. In that case, the absence of oceans on Venus was very hard to understand. Further, the large observed density differences between the terrestrial planets presented a serious problem for a model in which they all formed at nearly the same temperature out of the same materials.

Spectrophotometric and spectroscopic studies of the asteroid belt began in 1970 with the discovery that the large, bright asteroid Vesta, in the heart of the belt, had a reflection spectrum indistinguishable from that of laboratory samples of a class of igneous, differentiated (achondritic) meteorite.[12]

It was in this setting that I first proposed, in 1972, a model for the temperatures in the solar nebula at the time of formation of preplanetary solid materials.[13] In a detailed presentation of this model in 1974,[14,15] a number of specific interpretations and predictions were put forward. First, the high uncompressed density of Mercury was attributed to the accretion of Mercury out of material that had condensed and equilibrated in the solar nebula at such high temperatures that iron was mostly condensed, but the magnesian silicates were incompletely condensed. Second, the small apparent uncompressed density difference between Venus and earth, with earth actually slightly denser, was attributed to rather complete retention of the heavy volatile element sulfur by earth and its depletion in Venus. Third, the low uncompressed density of Mars was attributed to a higher degree of oxidation and hydration. The formation of Mars at lower temperatures than earth was a natural consequence of its greater distance from the sun; and a higher initial volatile content was an unavoidable consequence of lower formation temperatures. Fourth, the primitive material of the asteroid belt was predicted (Vesta nonwithstanding) to be volatile-rich, similar to the highly oxidized carbonaceous chondrites.

The close satellite systems of Jupiter and Saturn were interpreted as having formed in the presence of strong radial temperature gradients centered on the planets, in planetary subnebulae that emulated the structure of the solar nebula itself. The outer Jovian and Saturnian satellites formed at temperatures controlled by local conditions in the solar nebula, where highly volatile ice-forming solids were present in amounts that increased with heliocentric distance. Pluto, the subject of contemporary models that interpreted it as an "iron-rich" body with earthlike composition, was predicted to be the most ice-rich member of the sun's family of planets, and the lowest-density body orbiting the sun directly (except possibly some classes of comets, whose densities were and are unknown). The low density of Pluto has since been verified.[16]

The density of each "solid" planet was viewed as having been determined by the composition of material that condensed very close to that planet's location. The volatile element contents of the planets, however, can be strongly influenced and even dominated by small additions of material originating at much greater heliocentric distances.[17] A plausible accretion sampling model must accordingly be developed before definitive predictions of volatile contents could be made.

Discoveries in the planetary sciences have continued to accumulate since the early 1970s. I propose to conduct a brief tour of the planetary system, starting near the sun, with emphasis given to those observations that seem to contain information on the origins of the planets.

First, we shall consider Mercury, the innermost known body orbiting the sun. The high density bears witness to a metallic core with a mass equal to fully 60 percent of the mass of the planet, about twice as high a proportion as in Venus and earth. While high-temperature condensation in a very narrow temperature interval may be capable of producing local condensates with the requisite composition and density, reasonable accretion probability distributions would bring into Mercury so much material formed at lower temperatures that the observed high density becomes difficult to explain. Several special mechanisms have been proposed to account for Mercury's excessive density: very severe external bombardment that removed much of the crust and mantle; severe crushing of the brittle silicate component (and preservation of the malleable metallic component) during accretion; severe heating and boiling away of the crust and upper mantle by the superluminous phase of the early sun; and so on. While it is not possible to disprove any of these mechanisms, each has an ad hoc flavor, and would lead to disastrous and as yet unconsidered effects on Venus and the other planets. Aside from the dramatic enrichment of metal, very little is known of the composition and structure of Mercury. We believe that the planet is differentiated and has a core because all plausible thermal history models predict it, but no experiment has yet been carried out that is capable of establishing whether it is truly there. The surface composition is very poorly constrained by remote spectral data. A very faint FeO absorption band in the near infrared near 0.9 micrometers wavelength may be present, but the evidence is marginal. In any event, no more than a few percent of FeO could be present in the crust without providing clear spectral evidence for its presence.[18] The condensation theory outlined above suggests that the oxidation state (Fe oxide content) declines rapidly from earth, with about 10 percent FeO, through Venus to Mercury. The only spacecraft to fly by Mercury to date, the Mariner 10 mission, added nothing to our understanding of the composition of the surface rocks.

Venus has a massive atmosphere rich in carbon dioxide, with about 2 percent nitrogen. Radiogenic argon-40, made after the formation of the planet by the decay of the radionuclide potassium-40 in the crust, is similar in abundance to terrestrial radiogenic argon. The primordial rare gases differ greatly from those found in chondritic meteorites and on earth and Mars. First, the light primordial rare gases neon and argon (these are the non-radiogenic argon isotopes with weight 36 and 38) are roughly 100 times as abundant on Venus as on earth. The neon abundance is lower than that of argon, as in meteorites and the atmospheres of the other terrestrial planets, and not larger than argon, as in the unfractionated rare gases in the sun. No single effect seems plausible to account for these differences: an enhanced contribution of a solar-type rare gas component (to elevate the abundances of neon and argon relative to krypton, as observed) must be followed or modified by some other mechanism to deplete neon relative to argon in order to fit the observations of Venus.[19] After these two processes, the neon:argon ratio must mysteriously end up similar to that in ordinary chondrites, bodies upon which neither of these effects has operated!

The reactive minor gases, especially sulfur and the halogens, are present in the atmosphere of Venus in amounts so small that they could be supplied entirely from the observed infall rates of cometary and asteroidal material into the inner Solar System, even if past fluxes were no higher than they are at present. Such a conservative model also provides an amount of water roughly equal to the present atmospheric water content. More reasonable estimates of past cometary and asteroidal impact rates suggest that these elements plus nitrogen could have been supplied with ease by late infall of volatile-rich impactors, perhaps in 100-fold to 1,000-fold excess over their present amounts.[20] Only carbon dioxide is difficult to provide by such a mechanism, and therefore only carbon dioxide can be plausibly identified as a volatile component of Venus at the time of planetary formation. The interesting conjecture that comets are a major source of at least the lighter rare gases on Venus remains untested, and must await the results of mass-spectrometric analyses of the rare gases in cometary comas.

Several of the major reactive volatiles on Venus can form stable minerals under present Venus surface conditions.[21] The abundances of these solid carbonates, sulfates, sulfides, chlorides, fluorides and hydroxyl amphiboles are quite unknown, although some radar reflectivity data suggestive of volcanic sulfide deposits do exist.[22] Reactions between atmospheric gases and these minerals regulate, or "buffer," the abundances of their active gases at levels very close to their observed proportions.

Only one isotopic oddity is known among the chemically active volatiles on Venus: the present concentration of heavy hydrogen (deuterium; D) relative to normal hydrogen (H) is about 100 times as high as on earth or in chemically bound water in meteorites.[23] Simple models, in which Venus is given an initial endowment of ordinary water but loses H more rapidly than D due to the preferential escape of the lighter isotope from the planet, show that Venus must once have had at least 100 times as much water as is presently found in the atmosphere. This is nearly 1 percent of the amount of water in earth's oceans. More realistic models incorporating the effects of surface reservoirs of bound water and the powerful influence of cometary and meteoritic infall of water have yet to be developed: we do not in fact know whether the atmospheric water content on Venus is presently rising or falling. It is interesting to note that dynamical accumulation models for the terrestrial planets suggest that Venus should have accreted roughly 10 percent of its mass from earth's vicinity,[24,25] and hence Venus should have formed with a few percent of earth's water content.

Venus and earth are near-twins in several respects, but very dissimilar in others. Although their atmospheres are grossly different, the two planets are very similar in their distance from the sun, size, mass, density, surface gravity, and escape velocity. Mercury has a much higher density than either, and Mars has a much lower density than either. The uncompressed density, which is a crude but useful measure of composition, proves Venus to be rather similar to—but actually slightly less dense than—earth, thus breaking the general trend of decreasing density with increasing distance from the sun exhibited by the solid bodies that orbit the sun. This is actually a quantitative prediction of equilibrium condensation theory, which gives earth several times as much of the dense element sulfur. Note that sulfur (atomic mass 32) is much heavier than the prevalent element in the terrestrial planets, oxygen (mass 16), which dominates the mantles of the inner planets.[26]

Mars differs clearly from earth not only in density, but also in the distribution of density within the planet. Not only is the Martian core relatively less massive and less dense than the core of earth, but the Martian mantle is relatively more massive and denser than earth's. Equilibrium condensation theory attributes these differences to formation of Mars at lower temperatures, at which metallic iron is nearly fully oxidized to FeO and the volatile content of the planet is greatly enhanced. Less free metal means a core dominated by iron-nickel sulfides and a mantle rich in dense iron oxides. Higher volatile content suggests (but does not prove) massive and easy early outgassing.

The atmospheric composition of Mars is full of oddities. The primordial rare gases look remarkably similar in composition to terrestrial and chondritic gases, except that their absolute abundances (grams of gas per gram of planet) are about 100 times lower than on earth or in chondrites. Nitrogen is also very severely depleted relative to the same reference bodies. The surface of Mars bears powerful evidence of the effects of vast amounts of water and ice, and even the greatest of the canyons on its surface, which are closed and have no drainage outward onto the surface, may be karstic features caused by the extraction of underlying carbonate rock by acidic ground water. Insofar as water and carbon dioxide are concerned, it is easy to believe that Mars once was extremely volatile-rich, and even now harbors quite remarkable reserves of subsurface volatiles. But the rare gases and nitrogen seem to tell a different story.[27] The radiogenic isotope argon-40, made by decay of radioactive potassium-40 within Mars after planetary formation, is 3000 times as abundant as primordial argon on Mars (vs. 300 times on earth). Thus it is necessary to conclude that either:

   a. Mars formed deficient in many volatile elements,
   b. Mars is very inefficiently outgassed, or
   c. the early volatiles on Mars were somehow lost
      catastrophically.[28]

Of these, the first is least credible. Everything we know about the Solar System associates increasing distance from the sun with lower formation temperatures, higher oxidation state of iron, and higher volatile content. It seems pointless to throw out this powerful generalization without very good reason. The second can hardly stand alone: why should late-forming argon-40 be outgassed more efficiently than primordial argon? Until recently, the third possibility was dismissed out of hand because no mechanism for catastrophic (non-selective) lost of volatiles was known. More importantly, the idea smacked of catastrophism, a view of geology rejected two centuries ago in favor of uniformitarianism, the view that all the work of geology is done by the very slow, continuous operation of everyday processes like orogeny, weathering, transport, and sedimentation, operating over vast expanses of time. The crucial element for the success of uniformitarianism was that it had no need to postulate external (cosmic, celestial, divine) intervention in terrestrial affairs. The choice thus was made more on theological or philosophical than evidential grounds.

But planetary scientists understand that catastrophic processes of external origin are a natural part of the operation of nature. We have seen the impact scarring of Mercury, the moon, Mars and its satellites, and the satellites of the outer planets, and we have come home to earth with new eyes capable of discerning the badly weathered scars of equally titanic explosions here as well. Applying the knowledge of catastrophic bombardment processes won from the comparative study of the planets, we have recently come to realize that the intense impact bombardment associated with planetary accretion will almost certainly trigger massive volcanic outgassing and volatile release from severely shocked and disrupted impactors.[29] A growing planet will be surrounded by an envelope of freshly released gases. But the impacts occurring on Mars when grown to, say, 99 percent of its present size, will have a devastating effect on its atmosphere. Powerful shock waves from the impact explosion of a kilometer-sized asteroid or comet will blast a hole in the atmosphere, hurling up to 1 percent of the atmospheric mass off into space at speeds above the escape velocity of the planet. Such events are much rarer on earth because of earth's much stronger gravity: earth's atmosphere should accumulate without massive loss. Present crude models for the behavior of large ensembles of such impact events suggest that it is not difficult to blast away some 99 percent of its primordial atmosphere. The present atmosphere then is a combination of the last trace of released primordial volatiles with a contribution from radiogenic gases formed after planetary accretion.

The asteroid belt, between the orbits of Mars and Jupiter, is strongly zoned according to composition.[30] Based on recent spectrophotometric and spectroscopic data on over six hundred asteroids, we now know that a narrow ring at the inner edge of the asteroid belt is composed of stony material having the same basic minerals as the stony meteorites. Attempts to match the spectra of these stony (S-type) asteroids to laboratory spectra of ordinary chondritic meteorites, the most common classes of primitive meteoritic matter falling on earth, have been unsuccessful, and it is generally claimed by experts in this field that these bodies are more similar to stony-irons and achondrites (that is, igneous secondary materials) than to chondrites. The only asteroids that look like pieces of metal (M-class) are also present in this part of the belt. The heart of the belt is dominated by C-type asteroids, so named because of their strong spectral resemblance to laboratory samples of carbonaceous chondritic meteorites. Ceres, the largest asteroid, looks somewhat like an altered C-type. Its reflection spectrum contains strong bands due to chemically bound water in clay-type minerals (phyllosilicates). Among meteorites, the presence of such minerals is diagnostic of C chondrites. Beyond the C chondrites in the asteroid belt are the D and P classes, with spectra that look like "super-carbonaceous" variants of C chondrites, containing the same materials as the most volatile-rich chondrites, but in somewhat different proportions. Interestingly, Vesta, the first asteroid to be spectrally characterized, is now known to be highly atypical and possibly unique!

Beyond the orbits of the asteroids the only solid bodies we have observed are Pluto, the satellites of the outer planets and the possible escaped satellite Chiron. The satellite system of Jupiter shows a strong dependence of density (and volatile content) on distance from Jupiter: it seems certain that there was a strong radial temperature gradient in the Jovian subnebula analogous to that in the solar nebula itself. Callisto, the outermost of the large Galilean satellites of Jupiter, which was formed at temperatures close to those in the surrounding unperturbed solar nebula, is made mostly of water ice. The Saturnian satellite system is dominated by ices except perhaps in the region closest to the planet. The only other regular satellite system, that of Uranus, also appears to be dominated by ices. The Voyager 2 Uranus encounter should permit a vast increase in our knowledge of Uranus and its satellites.

The frontier of our planetary system, the realm of Pluto, its satellite Charon, and Neptune with its two strange satellites Nereid and Triton, is even less known, but all available evidence points to the domination of ices there as well. The atmospheres of Saturn's largest satellite, Titan, of Neptune's larger satellite, Triton, and of Pluto contain significant amounts of solid methane, a very low-temperature condensate.

Beyond the planets we have only the comets, which clearly are rich in water ice, other ices, and simple organic molecules. Their location(s) of origin are not well understood, but it is likely that they span a wide range of compositions and originate from throughout the expanse of space stretching from Jupiter to Neptune.

Systematic Trends and Conditions of Origin

By far the simplest interpretation of our present data on the nature and composition of Solar System bodies is that these bodies formed from vast ensembles of small solid planetesimals that had compositions that varied strongly with distance from the sun. The formation of the Jovian planets seems most readily explained by the accumulation of solid bodies with masses a few times that of the moon. These bodies then accreted nearby nebular gases gravitationally and ran away to very large masses.

All other Solar System bodies failed to become large enough for, or were in regions of the nebula that were too warm for, gravitational gas capture. These bodies perturbed each other gravitationally, collided, and accreted. As a result of gravitational interactions, the orbits of the small, unaccreted bodies were constantly being stirred up to higher eccentricities and inclinations. This assures that every small body will have a good chance of being perturbed into an orbit that crosses the orbit of at least one growing planet. Ultimately, a collision occurs, and the smaller body is accreted.

Thus every planet samples the surrounding population of small bodies in a statistical, rather than deterministic, way. A very small proportion of the mass accreted by a given planet will have been derived from very distant, volatile-rich regions of the nebula. In the case of the earth, the local "background" material was sufficiently volatile-rich that we cannot today discern with any confidence the signature of such a very-low-temperature (cometary) component. In the case of Venus, formed of much "drier" local material, and having so massive an atmosphere that explosive blowoff of gases is virtually impossible, the chances of discerning a cometary component are much better. It is only our vast ignorance of comets that prevents us from testing this idea today.

One apparently essential feature of planetary accretion is the coupling of outgassing with accretionary impacts. Students of the evolution of Mars have almost unanimously returned to the idea that Mars was formed of volatile-rich material, and the role of very large impacts in shaping the evolution of the terrestrial atmosphere is now being actively debated. It has become clear that secondary (post-accretion) events can affect the volatile element inventories of the planets, as was suggested many years ago by Fred Whipple. The quantitative assessment of these effects is now a burgeoning area of research.

A corollary of recent ideas on impacts is the realization that there must be an important stochastic component in the composition and mass of volatiles brought into a growing planet. The mass distributions of the early planetesimal swarms were so steep that the largest one or two impacts carried most of the mass. If, as suggested by spectroscopic studies, comets are indeed compositionally diverse, then the effects of the two largest impacts might be very different on two otherwise indistinguishable planets.

Future Prospects

Several parts of this puzzle may soon be at our disposal. First, several terrestrial spacecraft (from the European Space Agency, the

Soviet Union, and Japan) flew by Halley's comet in early 1986. An American spacecraft, designed for a wholly different mission, was dispatched from very high earth orbit to perform a flyby of yet another comet, Giacobini-Zinner, in the autumn of 1985. The famous Voyager 2 spacecraft, old and at least a little infirm years after its exciting encounters with Jupiter and Saturn, passed through Uranus's system in January 1986. It flew through the previously unvisited Neptune system in 1989 and provided crucial comparative data on the composition and thermal balance of Uranus and Neptune and on the chemical and physical properties of these planets' satellites and rings. Voyager, which was optimized for performance at the lighting levels and temperatures of the Jovian and Saturnian systems, was sent on to Uranus because the United States decided not to send a dedicated spacecraft optimized for the study of Uranus and Neptune.

There are no future spacecraft missions planned to visit Mercury. The USSR has announced the end of its extremely active and fruitful program of Venus exploration, and the U.S.A. will send a Venus Radar Mapper mission to examine the surface geology unimpeded by the dense, perpetual cloud cover. After a gap of fourteen years, both the U.S.A. and the USSR may be resuming flights of unmanned spacecraft to the moon. Both missions would be geophysical and geochemical mapping spacecraft in high-inclination orbits, capable of covering the entire lunar surface with their instruments and cameras. Both the European Space Agency (ESA) and Japan's NASDA have studies of similar lunar mapping missions under way.

Mars has again become the focus of renewed unmanned exploration. The U.S.A. will send a Mars Orbiter to map not only the surface geology and chemistry, but also atmospheric phenomena. In 1988 the USSR will fly two large spacecraft into orbit about Mars. Each of these vehicles will carry a small landing vehicle. Unlike earlier Soviet and American Mars missions, however, these probes will be dispatched to land on the surfaces of the Martian moons Phobos and Deimos, not Mars itself. In mid-1986 the Soviet Union announced two additional Mars missions to be launched during the 1990s. The first, tentatively scheduled for 1994, is a Mars rover mission loosely based on Soviet experience with lunar roving vehicles during the Lunokhod program. In 1998 the USSR plans to attempt an automated return of surface samples from Mars, again drawing upon their experiences with the Luna 16 family of lunar return vehicles.

There is only one future mission planned to explore the vast reaches of the outer solar system. That is the Galileo Jupiter orbiter and probe, which had been scheduled for launch by a Centaur G hydrogen-oxygen stage to be carried aloft by the Space Shuttle in May 1986. The loss of the Challenger orbiter has forced a two-year grounding of the shuttle. Also, one consequence of the thorough safety review of shuttle systems has been the decision to cancel the use of the Centaur stage on the shuttle. Several alternative methods for launching the Galileo spacecraft are presently under study, including the use of twin-IUS stages on the shuttle, and the use of conventional boosters such as the Titan 34D Centaur. Launch dates of 1989 to 1991 are presently under discussion. The spacecraft, when it is finally launched, will enter an orbit about Jupiter that takes it through a number of close encounters with the four huge Galilean satellites, and will drop its probe into the turbulent and brightly-hued atmosphere of Jupiter.

On its way to Jupiter, several months after launch, the Galileo spacecraft will fly by an asteroid and examine it with its powerful complement of instruments and cameras. The choice of the asteroid to visit depends sensitively upon the launch date and trajectory to Jupiter. In the early 1990s a Soviet spacecraft will be sent to carry out a landing on an asteroid. The actual target has not yet been chosen, but it is believed that a main-belt asteroid is preferred. Preliminary discussion has centered on Vesta. This mission would venture much farther from earth than any previous Soviet spacecraft. The Soviets have recently revised the plans for this mission to use a Mars (rather than Venus) swingby to assist the spacecraft on its way. Thus the Soviets have publicly acknowledged plans for four major sets of missions to Mars in the next twelve years.

The beginning of the era of exploration of comets, asteroids, and small satellites is the beginning of a new ability to examine ancient, primitive bodies, many of which are so small that they have been unable to evolve thermally and geologically. These are the most primitive, ancient, and unaltered objects left in the Solar System today. Their testimony to conditions during the time of formation of the planets will be more complete, more direct, less inferential than could be obtained from study of the larger, evolved bodies that has until now characterized our efforts in space exploration. We must expect that many of our favorite notions about the origin of the Solar System will need to be altered or rejected, and that our understanding will soon greatly surpass our present level.

It is never wise for scientists to issue dogmatic statements about truth without caveats appended: it is in the nature of scientific inquiry that we are constantly learning more, and constantly figuring out new ways to learn yet more. While most of our present ideas will probably find a comfortable home in the explanatory constructs of the next century, we may rest assured that there is much, much more to it than we presently know. The formation of a Solar System is necessarily an extremely complex chain of events, with many opportunities for variations. Chance may well have dealt us a few features that, on the cosmic scale of things, are quite rare. Likewise, certain phenomena that are very common in other planetary systems may by chance have been omitted or obscured during the assembly of our own. The true test of the validity of our ideas is not their ability to explain and predict the features of our own system, but to provide a broad and statistically reliable interpretation of the formation of planetary systems in general. By a curious coincidence of history, the maturation of our study of the Solar System with automated spacecraft is occurring simultaneously with the invention of a variety of new techniques to search for and characterize planetary systems of other stars. In the past year we have already seen the first pictures of dust disks forming into planets, and the first detection of a non-stellar "planetary" body in orbit around another star. The next few years will see the emergence of several of these powerful techniques, and with them will come the first information on the masses and orbits of planets in other stellar systems.

How common will other solar systems prove to be? Will about 20 percent of all stars have planetary systems, as I would currently guess, or is it 1 percent, 90 percent, or 0.001 percent? Will these planetary systems generally resemble our Solar System, with several rocky inner planets and several massive, gas-giant planets farther out, all orbiting in roughly the same plane? How massive is the biggest planet in each system—Jupiter-like, much smaller, or much larger? How variable are these general features from one system to the next? Within a few short years we should know the answers to many of these questions. We will then be in a position to ask a host of new questions that we presently know too little to formulate. But at the very least we will have begun to discern which of the features of our Solar System are mere local idiosyncrasies, and which are the cosmic norm. This distillation of the general from the particular is the true mission of science: now we shall at last be able to stand before Descartes, Kant, and the others who pioneered our craft and report to them how it really happened—and how unique we really are.

NOTES

1. A. G. W. Cameron and M. R. Pine. *Icarus* 18 (1973): 377.
2. P. Goldreich and W. R. Ward. *Astrophys. J.* 183 (1973): 1051.
3. G. W. Weatherill. *Geochim, Cosmochim. Acta Suppl.* 6 (1975): 1539.
4. V. S. Safronov. *Evolution of the Protoplanetary Cloud and Formation of the Earth and Planets.* Tel Aviv: Israel Program for Scientific Translations, 1972.
5. S. J. Weidenschilling. *Icarus* 27 (1976): 161.
6. J. S. Lewis. *Science* 186 (1974): 440.
7. J. S. Lewis. *Icarus* 16 (1972): 241.
8. J. S. Lewis and R. G. Prinn. *Planets and Their Atmospheres—Origin and Evolution.* New York: Academic Press, 1984.
9. D. J. Stevenson. *Planet. Space Sci.* 30 (1982): 755.
10. R. T. Dodd. *Meteorites.* Cambridge: Cambridge University Press, 1981.
11. E. Anders. *Accounts Chem. Res.* 1 (1968): 289.
12. T. B. McCord, J. B. Adams, and T. V. Johnson. *Science* 168 (1970): 1445.
13. J. W. Lewis. *Earth Planet. Sci. Letters* 15 (1972): 286.
14. J. S. Lewis. *Science* 186 (1974): 440.
15. J. S. Lewis. *Sci. Amer.* 230, no. 3 (1974): 50.
16. M. J. Lupo and J. S. Lewis. *Icarus* 44 (1980): 41.
17. See note 14.
18. F. Vilas and T. B. McCord. *Icarus* 28 (1976) 593.
19. G. W. Weatherill. *Icarus* 46 (1981): 70.

20. G. W. Weatherill. *Geochim. Cosmochim. Acta Suppl.* 14 (1980): 1239.
21. J. S. Lewis. *Earth Planet. Sci. Letters* 11 (1971): 130.
22. P. G. Ford and G. H. Pettengill. Science 220 (1983): 1379.
23. T. M. Donahue, et al. *Science* 216 (1982): 630.
24. See note 3.
25. See note 5.
26. See note 14.
27. T. Owen, et al. *Science* 194 (1976): 1293.
28. E. Anders and T. Owen. *Science* 198 (1977): 453.
29. G. H. Watkins. Ph.D. dissertation, M.I.T., 1983.
30. J. Gradie and E. Tedesco. *Science* 216 (1982): 1405.
31. *Planetary Exploration Through Year 2000.* Washington, D.C.: Solar System Exploration Committee, 1983.

Figure A.—This color rendition of Mars was made from three separate pictures taken through color filters—violet, green, and red—on June 17 as Viking 1 closed to within 560,000 km (348,000 miles) of the planet. The triplet of black-and-white frames was shuttered just seconds apart by one of the Viking Orbiter's two TV cameras. Corrections were made in the computer for the color response of the camera to reconstruct the photograph, which shows Mars as it would appear to someone approaching the planet. Clearly seen are the Tharsis Mountains, the row of three huge volcanoes standing about 20 kilometers (12½ miles) above the surrounding plain. Olympus Mons, Mars's largest volcano, is toward top of picture. North is toward upper right corner of photo. The circular feature at bottom of disk is a large impact basin, Argyre. The area around Argyre is slightly brighter than elsewhere, probably because of the presence of discontinuous thin carbon dioxide ice on the surface. Several atmospheric features are faintly visible. To the west of the southernmost Tharsis volcano (left) is an irregular white area, which has been seen on two successive days by Viking 1 and has been interpreted as a water-ice surface frost or ground fog. Faint, light, curved bands in the lower half of the picture probably are thin cirrus-like clouds. The yellows toward the edge of the planet's limb are somewhat artificial and are caused by extreme variation of brightness in the violet. (Photo: NASA)

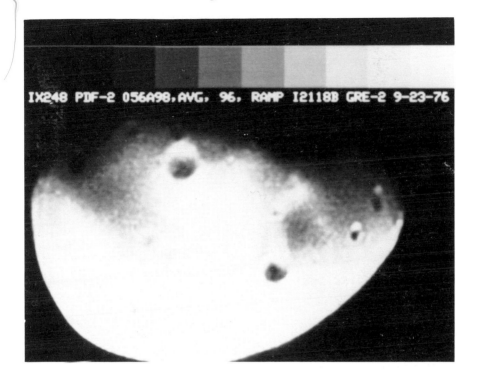

Figure B.—This is a computer-generated color picture of Deimos, smaller of the two satellites of Mars. A pair of images of Deimos from Viking Orbiter 1—one taken through the camera's violet filter, the other through the orange filter—were combined in this single image to search for color differences on the surface of Deimos. Resolution in this picture shows objects as small as 200 meters. Deimos is a uniform gray color: slight tints of orange on the rim of some craters are artifacts of the image process. A small blur beside the large crater at right is where scientists removed a reseau mark from the original image. The reseau marks are etched on the imaging system and are used to make precise measurements of the objects in the photos. (Photo: NASA)

All photographs on the following pages are provided courtesy of Jet Propulsion Labs. The photos were taken by the Voyager spacecraft.

Jupiter with moons Io (left), Europa (below Jupiter), Ganymede
 (lower left), Callisto (lower right).

TOP: Jupiter from equator to southern polar latitudes close to
 Great Red Spot.
BOTTOM: Io from 77,100 miles.

Saturn and three moons, Tethys, Dione and Rhea, Aug. 4, 1981.
13 million miles.

TOP: Saturn's rings. Photo Nov.12, 1980. Range 717,000 km.
BOTTOM: Saturn's moon Titan and thick haze. Photo Nov. 12,
 1980. Range 435,000 km.

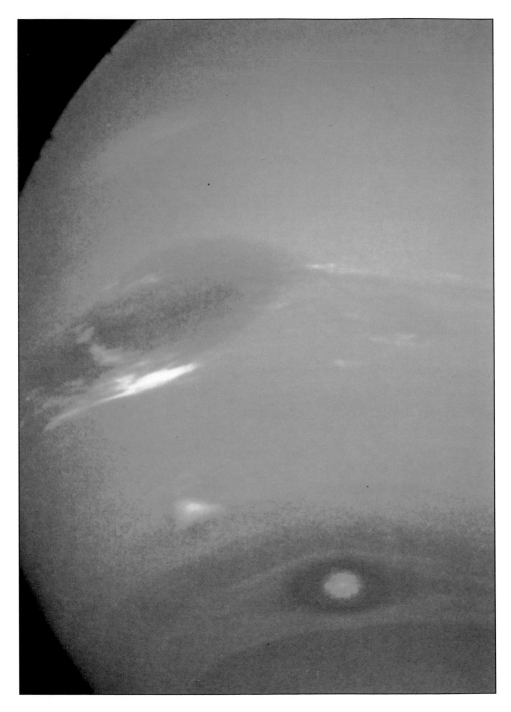

Neptune's Great Dark Spot, accompanied by white high-altitude
 clouds.

Farewell shot of crescent Uranus as Voyager 2 departs.
January 25, 1986. Range 600,000 miles.

Figure C.—Jupiter, its Great Red Spot, and three of its four largest satellites are visible in this photo taken February 5, 1979, by Voyager 1. The spacecraft was 28.4 million kilometers (17.5 million miles) from the planet at the time. The innermost large satellite, Io, can be seen against Jupiter's disk. Io is distinguished by its bright, brown-yellow surface. To the right of Jupiter is the satellite Europa, also very bright but with fainter surface markings. The darkest satellite, Callisto (still nearly twice as bright as earth's moon), is barely visible at the bottom left of the picture. Callisto shows a bright patch in its northern hemisphere. All three orbit Jupiter in the equatorial plane, and appear in their present position because Voyager is above the plane. All three satellites show the same face to Jupiter always—just as earth's moon always shows us the same face. In this photo we see the side of the satellites that always face away from the planet. Jupiter's colorfully banded atmosphere displays complex patterns highlighted by the Great Red Spot, a large, circulating atmospheric disturbance. This photo was assembled from three black-and-white negatives by the Image Processing Lab at Jet Propulsion Laboratory. JPL manages and controls the Voyager project for NASA's Office of Space Science. (Photo: NASA)

Figure D.—Saturn's rings are bright and its northern hemisphere defined by bright features as NASA's Voyager 2 approaches Saturn, which it encountered on August 25, 1981. Three images, taken through ultraviolet, violet, and green filters on July 12, 1981, were combined to make this photograph. Several changes are apparent in Saturn's atmosphere since Voyager 1's November 1980 encounter, and the planet's rings have brightened considerably due to the higher sun angle. Voyager 2 was 43 million kilometers (27 million miles) from Saturn when it took this photograph. The Voyager project is managed by the Jet Propulsion Laboratory, Pasadena, California. (Photo: NASA)

13

Structures of the Terrestrial Planets

R. A. LYTTLETON

The origin of the earth and other planets of the solar system poses one of the great central problems of astronomy. The famous nebular hypothesis of Laplace,[1] which dominated ideas for nearly a century, considered all the planets to be initially gaseous. It provided no origin for the nebula itself, but the ultimate central body to which the nebula contracted would have had far too rapid rotation to be identified with the sun. It may be noted even at this point that the moon and Mercury can hold no appreciable atmosphere now, and Mars barely so, and they can never therefore have been in all-gaseous form, as in fact also turns out to hold true for earth and Venus. The sun being central to the system led theorists to turn to tidal or collisional encounters with a passing star, but this also implied initially gaseous planets, and also failed by a very large factor to account for the enormous scale of the system compared with the sun—Jupiter moves at a distance of 1,000 solar radii and Neptune at 6,000. This difficulty was overcome by the hypothesis[2] that the sun originally possessed a companion star moving at a distance of planetary order from it, and by means of a close encounter of this body with a passing star, the incipient planetary material would now be released. The encounter itself could speed up the companion star sufficiently to break its weak gravitational binding to the sun and cause it to escape. In a variation of this binary theory, the companion was supposed to be sufficiently massive to have evolved rapidly to a supernova-explosion stage, the sun capturing a mere wisp of the vast amount of material (several sun-masses) thereby thrown off into space, the slightest asymmetry of recoil enabling the remnant star to be set free from the sun. The captured gaseous material would expand and cool, some into dust, and settle into a pre-planetary nebular disc in rotation round the sun, and this material would proceed to develop into a set of planets. These processes of formation would imply that other planetary systems than our own would be extremely rare.

Up to this stage, interstellar space had always been regarded as practically empty, apart from a few calcium atoms, but the situation was completely changed by the steady discovery of vast quantities of gas clouds and dust clouds, mainly in association, that go to constitute the main form of the galaxy and occupy about 15 percent of its overall volume. Such clouds have dimensions of parsecs and masses several times that of the sun, and pursue orbits in the galaxy just as do the stars, so that in addition to the huge general galactic rotational velocity they have superposed more or less random velocities of a few kms/sec relative to adjacent objects—stars and clouds. Now it is dynamically possible for a star such as the sun to acquire material from one of these gas-and-dust clouds. As our planets have total mass only $1/700$th that of the sun, it is evident that to provide for such an amount the sun would need to have captured only the merest wisp of material from such a cloud. Capture could easily result through intervention of another comparably near star in slowing just part of the cloud relative to the sun. Such a star would not have to come unusually close to the sun for the effect to operate, since the clouds themselves have dimensions of interstellar order. The exact amount of material captured and its angular momentum about the sun, which latter is what would determine the eventual scale of the resulting system, would depend on more or less chance factors. But the interesting point now emerges that planetary systems so formed may be extremely numerous in the galaxy, and other galaxies, though usually with orbital scale and total mass different from our own, and each with a variety of planets physically as in our own system. Similar dynamical action would lead to planetary systems in the very numerous external galaxies that are seen to contain gas-and-dust clouds, the latter component in particular being essential for production of planets of terrestrial type.

There is a second, more direct way that a star such as the sun would collect interstellar gas and dust without the action of another star, and this is by the process of accretion through gravitational focusing of material towards the axial line behind the sun as it passed through the cloud.[3,4] At one extreme, if the star were at rest within such a cloud, its attraction would draw in material radially, but if the cloud had vorticity or rotary momentum as such clouds do, this would be conserved and the cloud could contract down only to an extent limited by centrifugal force. At the other extreme, if the star moved through the cloud at very high speed, it would collect little more than just the material it ran into bodily, and the mass captured would be negligibly small and have no orbital motion. However, somewhere between these extremes exists a relative speed that would allow capture of a mass comparable with the mass of the planets.

Depending on the density of the cloud, the requisite speed is of order 1 km/sec, and this is small compared with the relative velocities of stars and clouds, which have distributions averaging about 20 km/sec. But in the accretion process, the star interacts with the *whole* cloud, not merely with the little it captures, and this results in a strong braking action, reducing its speed relative to the cloud, with the rate of capture increasing as the inverse cube of the speed. Thus an initial speed such as 5 km/sec could well be slowed almost to zero by this braking action during passage of a star through a large cloud and hugely increase the total capture. In this way, perhaps one star in a hundred has undergone an encounter enabling capture of a mass of planetary order, later to aggregate into planets, but even this could mean that 10^9 planetary systems are associated with stars in our galaxy. That such a process frequently is undergone by the sun is shown by the existence of myriads of comets, of which several million are in orbit abut the sun at present. There are of the order of a hundred groups of these that come in towards the sun from almost identical points of the celestial sphere for each group. The groups are widely spread over the whole sphere, but show a strong belt-like distribution parallel to the galactic plane and also concentration towards the ant-apex of the motion of the sun. The comets are vast swarms of widely spaced dust-particles forming in passages of the sun through dust-clouds at higher speeds, though still below average, and only a small amount of material compared with the planets is captured at each such encounter with a dust-cloud.

As said earlier, these galactic clouds are in slow rotation sufficient to prevent them collapsing under their self-gravitation, which means angular velocities of a few time 10^{-15} per sec. Even this seemingly minute rate implies considerable angular momentum for the captured material, and it may be shown to be of the same order as that carried by the orbital motions of the planets. But there is possibly a stronger source of such momentum, for not only are clouds quite irregular in form and density, and but a star would be most unlikely to pass symmetrically through one. The amounts of material converging towards the accretion axis from opposite sides of some plane through it would usually be quite unequal, and a difference as small as 1 per cent would be adequate to account for the planetary angular momentum, and could be more effective than any intrinsic vorticity in the cloud.[5]

Once captured by the sun, the development of such a gas-dust nebula is readily determined. The 1 or 2 percent by mass of its dust particles will rapidly take up the form of a flat annular disc as thin as possible with its areal elements all in Keplerian orbits round the Sun.[6] This is easily proved, but it is actually exemplified in nature by Saturn's rings whose particles are confined to circle the planet in an extremely thin plane annulus. The particles cannot coagulate to form moons of Saturn as they are closer to the planet than the Roche limit at which distance differential orbital rotation overcomes their mutual gravitation. However, this is by no means so for the planetary dust disc at far greater proportional distance from the sun, and which would have flattened to a thickness of only a few centimeters at the distances where the terrestrial planets would then come to form. Incipient planets can grow by gravitation from chance accumulations within this disc and progressively increase in mass till all the material of the disc has been absorbed into them to produce the primitive terrestrial planets. By such means the earth could gradually grow to practically its total mass in about 5×10^4 years. Moreover, growth of a planet by such means from a thin disc has two highly important consequences, as will now be explained.

First, captured material would necessarily fall to its surface only in a very narrow circumferential band just a few centimeters wide and not to all parts of the surface. When the mass becomes sufficiently great for the infalling impact with the surface to produce melting, which would occur for an object the size of the moon or greater, the material thereby liquefied would not pile up where it fell but flow sideways from the narrow band in polar directions. This spreading on the surface would be especially conducive to rapid cooling of the material to solid form, and thus even the earth and Venus would originate in all-solid form, as has already been seen, the moon, Mercury, and Mars must have so originated.

Second, because of their orbital motions, small areal elements of the disc will possess vorticity with axis perpendicular to the plane of the disc; as mass is drawn into a growing planet the vorticity is conserved, and thereby rotation is imparted to the body. Planets formed in this way would initially have rotation periods of order four to five hours about parallel axes. This close parallelism still holds for a large proportion of planetary mass, which is mainly carried by Jupiter and Saturn. The exceptions to this rule can be accounted for by subsequent large additions of mass, and by tidal and dynamical action. There is also evidence from the excessive radioactive content of the outer layers of the earth, also found on the moon, that a second capture of interstellar material of such nature by the sun took place, gradually swept up onto the already existing planets. If the initial plane of the disc of this material, which would have had mass of an order of 1 percent of the planets, is tilted its angular momentum would disturb the axes of rotation of the planets by amounts of the order of a degree or so, as presently exhibited for the most part.

At the distances of the earth and other terrestrial planets, the solar-maintained temperature would be too high for hydrogen, the main interstellar constituent captured, to be collected by such small planets. Even farther out, at the distance of Jupiter and beyond, there would first be needed aggregation of a solid planetary core of mass comparable with that of the earth. At the lower temperature there prevailing, this core could then go on to retain hydrogen and other light gases, and gradually accrete all of the primal captured gases. This explains why the great planets not only have masses some two

orders of magnitude greater than the inner planets but mean densities only of the order of unity (1g cm^{-3}) because of their high content
of hydrogen, compared with about 3–5 g cm^{-3} for the terrestrial
planets.

A further important consequence for the inner group of planets
from a well-mixed dust cloud is that they must all have similar
chemical composition. There could be no possibility, for example, of
the earth first forming an iron core and then going on to surround
this by purely rocky material, as is often supposed to be its structure.[6] The well-established seismic data for the earth reveal all its
physical properties at every depth, apart from the temperature and
chemical composition, and these known elastic properties can be
applied to the earlier forms of the earth and to the present and earlier
forms of the other inner planets and the moon, and fully account for
their principal features and interior mass distributions. Thus when
the earth was all-solid, in which form it remained for about 10^9 years,
its radius would have been about 370 kms greater than at present. But
as a result of internal radioactive heating there occurred intermittent
epochs of rapid overall contraction, at times of the order of 10^8 years
apart, which occasions correspond to eras of intense mountain-
building. There are known to have been more than twenty such eras
during the age of the Earth.[7,8]

It is also possible using the seismic data to calculate the initial
moment of inertia of the earth; when the crucial stage of mountain
building was first started about 3×10^9 years ago, the value would
have been 9.78×10^{44} cgs units compared with the present value of
8.15×10^{44} cgs. It follows that the average rate of decrease during the
past three aeons has been $\mathbb{C} = -1.70 \times 10^{27}$ cgs. Now dynamical analysis of the ancient eclipse data, which as yet cover only the past 3,500
years, inescapably requires a changing moment of inertia, and the
rate emerging from the most accurate values of the secular accelerations of the moon and sun is $\mathbb{C} = -1.67 \times 10^{27}$ cgs, a value in remarkably close agreement with the average rate during the whole life of
the earth.

This theoretical approach to planetary formation enabled it to be
predicted, before any space missions to planets, that unlike the earth
there would be found no folded and thrusted mountains on the moon,
Mercury, and Mars, but that such mountains would be found on
Venus. All these predictions have now been proved correct. Much has
been made of the longstanding general acceptance of a high density
for Mercury, widely believed to be almost as high as that of the earth
at a figure of 5.4 g cm^{-3}, implying an iron content of about 60
percent. This result rests ultimately on a value promulgated by Newcomb in 1895, but investigation of his work has revealed that after an
extensive study over a period of several years, which gave a closely
accurate and acceptable value for the mass of Venus, he abandoned
its implications for the mass of Mercury and quite unjustifiably
adopted on no valid basis a value of sun/6,000,000, when his actual

arithmetic if followed out would have led to a mass of Sun/8,500,000, which latter value would bring the density and hence the composition of the planet into line with the other terrestrial planets and the moon.[9] There is urgent need for thorough re-discussion of the Mercury mass that can in fact now be accurately determined far more straightforwardly and precisely from the Viking data than by any other means.

REFERENCES

(1) Laplace P. S., Exposition du Systeme du Monde, 1796.
(2) Lyttleton R. A., The Double-Star Hypothesis, Monthly Notices of the Royal Astronomical Society, *98*, 537, 633, & 646, 1938.
(3) Bondi H. & F. Hoyle, On the Mechanism of Accretion by Stars, Mon. Not. Roy. Astron. Soc. *104*, 21, 1944.
(4) Lyttleton R. A., A New Solution to the Accretion Problem, Mon. Not. Roy. Astron. Soc. *160*, 255, 1972.
(5) Lyttleton R. A., The Formation of Planets from a Solar Nebula, Mon. Not. Roy. Astron. Soc. *158*, 463, 1972.
(6) Lyttleton R. A., The Earth and Its Mountains, John Wiley & Sons, 1982.
(7) Holmes A., Principles of Physical Geology, Ronald Press, 1945.
(8) Leet L. D. & Judson S., Physical Geology, Prentice Hall, 1961.
(9) Newcomb St., The Four Inner Planets & the Fundamental Constants of Astronomy, Washington, 1895.

GIANT COMETS— THEIR ROLE IN HISTORY

14

Giant Comets and Their Role in History

S. V. M. CLUBE

Several discoveries during the last decade indicate that a disintegrating giant comet (> 100 kilometers) has probably dominated the terrestrial environment during recent millenia. The destructive effects of its smaller debris (< 1 kilometer in size) on the earth typify physical processes that have apparently been going on with greater or lesser intensity throughout the geological past, thereby controlling terrestrial evolution. The significance of such debris has been overlooked in conventional astrophysical theory, so it is likely that this new understanding of earth history will have far-reaching consequences in fundamental areas of physics and astrophysics.

Depending on one's viewpoint however, the consequences may be even more significant for our understanding of the growth of civilization. Thus it is very probable that the continuing breakup of the most recent giant comet and its calamitous consequences were witnessed and experienced by our ancestors, particularly between two and one hundred generations ago, indelibly imprinting the firm conclusion that humanity was at the mercy of a visible, purposive, divine being in the sky. Although many twentieth century human activities are still unwittingly justified by theologies that derive from this conclusion, western civilization effectively abandoned the conclusion during the Enlightenment when ancient knowledge of the witnessed effects (e.g., myths) seemed no longer to have any meaning. It follows that the new discoveries introduce for the first time the possibility of forging a link between modern science and ancient knowledge, and furthermore of opening our eyes to long-term hazards that seriously affect our terrestrial environment (i.e., cometary winters, nature's version of a nuclear winter). The exact nature of the most immediate celestial hazard remains to be discovered but such an enterprise, not yet undertaken by science, would appear to be as important as any plan to control this planet or colonize the others!

The Question of Comets

In the West a star shall shine, which they call a comet, a messenger to men of the sword, famine and death.

The Sibylline Oracles

God, whose dwelling is in the sky, shall roll up the heaven as a book is rolled, and the whole firmament in its varied forms shall fall on the divine earth and on the sea; and then shall flow a ceaseless cataract of raging fire and shall burn land and sea, and the firmament of heaven and the stars and creation itself it shall cast into one molten mass and clean dissolve. Then no more shall there be luminaries, twinkling orbs, no night, no dawn . . . no spring, no summer, no winter, no autumn.

The Sibylline Oracles

"You [Greeks] are all young in your minds," said the priest, "which hold no store of old belief based on long tradition, no knowledge hoary with age. The reason is this. There have been, and will be hereafter, many and divers destructions of mankind, the greatest by fire and water, though other lesser ones are due to countless other causes. Thus the story current also in your part of the world, that Phaeton, child of the Sun, once harnessed his father's chariot but could not guide it on his father's course and so burnt up everything on the face of the earth and was himself consumed by the thunderbolt—this legend has the air of a fable; but the truth behind it is a deviation of the bodies that revolve in heaven around the earth and a destruction, occurring at long intervals, of things on earth by a great conflagration. . . . Any great or noble achievement or otherwise exceptional event that has come to pass, either in your parts or here or in any place of which we have tidings, has been written down for ages past in records that are preserved in our temples [in Egypt]; whereas with you and other peoples again and again, life [had only just] been enriched with letters and all the other necessities of civilization when once more, after the usual period of years, the torrents from heaven [swept] down like a pestilence, leaving only the rude and unlettered among you. And so you start again like children, knowing nothing of what existed in ancient times here or in your own country. . . . To begin with, your people remember only one deluge, though there were many earlier; and moreover you do not know that the noblest and bravest race in the world once lived in your own country. From a small remnant of their seed you and all your fellow citizens are derived; but you know nothing of it because the survivors for many generations died leaving no word in writing. . . ."

Timaeus (Plato, tr. F. M. Cornford)

> We still tremble today from the consequences of the deluge, and our institutions, without our knowing it, still pass on to us the fears and the apocalyptic ideas of our forefathers. Terror subsists from race to race, and the experience of the centuries can only weaken it but cannot make it entirely disappear.
>
> *L'Antiquite devoilee par ses usages* (Boulanger, 1766)

The earliest natural philosophers of ancient Greece[1] developed a cosmogony in which the world came into being through the production and condensation of four elements (fire, air, water, earth) from an original undifferentiated mass or "cosmic egg." The world formed out of these elements was basically a temporary flat system embracing the sun, moon, and earth. A succession of such worlds was apparently envisaged, each manifestation being in due course destroyed and returned to the so-called "boundless." An important but commonly overlooked feature was the presence of "stars" *below* the sun and moon. These "stars" seem to have been pictured as luminous jets emerging from orifices that moved around otherwise enclosed tubes of compressed fire bent into the shape of hoops. This particular imagery is evidently well suited to a comet in a short-period orbit and indeed, it should be remembered that "stars" in earlier times generally were often synonymous with comets. As recently as the fourteenth century, for example, Giotto depicted the star of Bethlehem by a comet while another well-known apparition, that of Comet Halley in 1066, is also described as a star on the Bayeux tapestry. Thus, the early Greek accounts, in which a cosmic egg constructs a temporary flat world, may be looked upon as attempts to describe the process by which an active short-period comet enhances the zodiacal cloud. In fact, there are several indications that the earliest references to the Milky Way are descriptions of an intense zodiacal cloud and that it, too, was able to reach below the moon.

The planets are usually not referred to in these early accounts; indeed it is probable that they were unimportant and that the major concern of pre-Socratic philosophers was with meteoric phenomena. The Greek tripartite classification of comets into "bearded," "cypress tree," and "torches," of which several kinds of the latter are clearly descriptions of bolides or even meteorite falls, lends support to this view. Thus, although the attention to meteoric phenomena in classical times has appeared somewhat anomalous to many commentators, an enhanced fireball and meteoritic flux associated with the evolution of a large disintegrating comet is not at all beyond the bounds of possibility.

Sometime around the fourth century B.C., however, cosmological theorizing in ancient Greece underwent a profound change. A completely new tradition of planetary observation emerged, together with a growing academic interest in geometry and circular motion. Originating apparently with the Pythagoreans and gaining momentum in the hands of Eudoxus and Plato, this new tradition gave birth to the highly radical Aristotelian cosmology in which comets completely lost their status as celestial objects, together with their portentous overtones. Even Aristotle admitted, however, that comets were a source of dust replenishing the Milky Way, and that one may frequently observe more than one comet at a time! Thus, it seems not improbable that the Aristotelian paradigm shift took place during a period of declining comet activity and diminishing zodiacal cloud, both of which were nevertheless still enhanced above current levels. Now, it has been argued that Aristotelian cosmology was a largely aristocratic construct motivated by the practical needs of empire; and it is of course possible that by defusing catastrophic celestial missiles, the destabilizing influence of public alarm could be usefully moderated. Plato's reference to bodies that revolve in heaven around the earth and that bring destruction at long intervals by a great conflagration would then also have been relegated to the level of myth. Now, it is known that the scientific purpose of the Aristotelian system in the hands of the Alexandrian school was merely to "save the phenomena," and that it also had theological connotations that, if Strato is not misunderstood, were attributable to a wholly invented divine control. One can appreciate therefore why, in addition to an immobile earth and its concentric spheres supporting the moon, Mercury, Venus, the sun, Mars, Jupiter, Saturn, the fixed stars, and the invisible Prime Mover, it also became a fundamental feature of the Aristotelian tradition to confer divine status upon the planets and to attribute human behavior to the remote action of these new gods, describable in terms of a quite imaginary theoretical structure that is now known as "horoscopic astrology." This scheme found its highest expression in Ptolemy's works, the *Almagest* and the *Tetrabyblos*, and it continued to appeal to the leading minds of Western civilization for most of the subsequent fifteen hundred years.

Since this time, there has been a tendency to see the growth of science as an exclusively mathematical tradition; and commentators given to this viewpoint tend to treat the *Tetrabyblos*, quite wrongly, as an aberration outside the mainstream of science. The mathematical tradition is very much the domain of leading minds and it may have been unpalatable to see it as a mere response to a perceived physical framework. In the event, it needed a Descartes to detect the fundamentally irrational character of astrological forces "acting at a distance," and to expose the capacity for self-delusion on the part of natural philosophers for the sake of a wholly contrived and unsupported theory.

Although Aristotelian theory remained in the ascendancy for nearly two thousand years, the population at large was never entirely convinced as to the unimportance of comets and continued to regard them with considerable awe. It was natural, therefore, that medieval astronomers should pay particular attention to comets once the fallacies of Aristotelian cosmology were exposed, and it was Tycho Brahe who eventually introduced a radical change of view. Thus by making careful observations of the comet of 1577, he discovered that its nearest point of approach was well outside the sublunary zone. Conceivably then, comets were instruments of divine vengeance after all, but those of the time who were less eschatologically inclined took a more prosaic view. Kepler for example, simply assumed comets were sweeping past the Solar System in straight lines, while Galileo, anxious perhaps to be rid of all the humbug, treated them as mere optical illusions! The latter was not a particularly tenable viewpoint, however, so we find Newton subsequently preferring the Keplerian assumption, though not so much out of conservatism as it happens, but because he wished to avoid comet encounters doing any damage to his perception of the Solar System. Thus Newton, like Aristotle before him, considered the Solar System to be a divine creation, set to run like a clockwork machine; and stray comets passing through could obviously put it very much at risk. In 1681, however, another important comet was observed, this time by Flamsteed, and it was without doubt in a near-parabolic orbit, coming well within the planetary system. Never slow to come to terms with the inevitable, Newton had to rapidly adjust his stance—the orbits of comets were obviously now just another aspect of the divinely ordained law of gravity. But with a subtle change of emphasis, Newton insisted that encounters with planets should now be seen as providential rather than catastrophic events. Very little attention has been given to this interesting circumvention of Newton's, but so far as its impact on science is concerned, it has been almost as significant as his law of gravity. Let us follow the matter a little further.

Newton's Other Law

Newton was, of course, not alone in reflecting on these questions; and by this time the possibility of a disaster through cometary impact on the earth had become a matter for general debate.[2] Scholars seriously questioned whether the conflagration of Phaeton and the flood of Deucalion or Ogygus had been caused by a celestial body. The flood of Ogygus was noted, for example, as having been attributed to the arrival of a comet Typhon, and there was also a connection with the Bible, for Roman historians had made the plagues of Egypt a contemporary event. Such issues as whether the floods of Noah, Ogygus, and Deucalion were one and the same were also raised, and the outcome of all these inquiries was a growing concern for terrestrial catastrophism and ancient chronology. On the one hand, it seemed that historical studies might help to determine the periods of returning comets, and on the other hand, it seemed that one could use such periodicity to plot the course of major historical events.

Then, by chance, another comet appeared in 1682, and Halley was one of the first to identify it with recent well-authenticated apparitions in 1531 and 1607, thus providing a demonstration of the methods to be applied. This particular comet was of course to take Halley's name in the fullness of time (following its predicted return in 1758) but Halley's more immediate interest was in deriving a period for the comet of 1681. A value of 575 years was obtained (incorrect as it happens), which Whiston, Newton's successor in the Lucasian chair at Cambridge, made use of in calculating earlier apparitions. Notable among these was one at Caesar's death in 44 B.C. and another corresponding to Noah's flood in 2342 B.C. Indeed, Whiston's book describing these researches, entitled *The New Theory of the Earth* was dedicated to Newton. The book came rapidly to be seen by clerics and natural philosophers alike as a quite major step forward in the advancement of knowledge. Seemingly for the very first time, observational science and received biblical knowledge were in excellent accord. It was a history of catastrophe moreover, and the day of judgment could even be at hand!

Whiston, however, did not see it this way. He thought the course of events required a more matter-of-fact interpretation of the Bible, and that it indicated a less prominent role for Christ. This was taken by his contemporaries to be an attempted revival of the old Arian heresy, and the clerical establishment soon sought to distance itself from Whiston. In the meantime, Newton took it upon himself to emphasize what he considered to be the fundamental role of comets, namely their ability to deposit new material on the stars and planets and their potential for doing good to the earth! The calm voice of reason thus spoke. Scaremongering was naturally added to the list of Whiston's faults, and it was not long before the latter was dismissed from his post. Even if there was no conspiracy, the message from the lawmakers was clear: the world was not to be disturbed and Newton's new ground rules for comets were the ones that had to be observed. The return of Halley's comet in due course, in accord with gravitational law, then merely served to enforce the new paradigm—as also did Lexell's comet, some years later, when it passed by the earth without any noticeable gravitational effect. The absence of the effect was good evidence that comets were very much smaller than planets, thereby diminishing still further any concern that still lingered concerning the menace that comets seemed to present. Thus, by the end of the eighteenth century, Newton's view of comets had gained considerably in strength, and we find Herschel, a very influential astronomer, putting forward an almost teleological view of comets as a necessary adjunct to his new discoveries regarding the distribution of stars in the Milky Way. It was Herschel's opinion, for example, that comets were to be taken as essentially interstellar objects purposely weaving their way among the stars, including the sun, with the specific intention of replenishing their fuel, and if necessary, planetary life as well. Newton's theory had therefore developed rather successfully into an all-embracing cosmological view of some permanence and continuity to which the masters of an orderly empire could easily subscribe. It was a vision that clearly deserved royal patronage!

The new discoveries of meteorites and minor planets that happened to come at this juncture might have seriously disturbed the status quo by raising the spectre of other kinds of celestial hazard, and they were indeed ferociously opposed; but the generally small size of the former and the confinement of the latter to a belt between the orbits of Mars and Jupiter eventually allayed any fears. Thus there was little doubt in anyone's mind that the natural philosophers contributing to the Enlightenment had been broadly right in considering only the effects of comets. Even if by the middle of the nineteenth century some questions were being raised concerning Herschel's view of the interstellar nature of comets—because of the failure to observe any on hyperbolic orbits—the by now established view of catastrophism was being upheld in its essentials by the growing realization that comets trapped into short-period orbits rapidly decayed into meteor streams. It had indeed come to be appreciated that nature was capable of rendering comets more or less automatically harmless by turning them into dust that fed into the atmosphere to be witnessed as shooting stars. Thus, well before the middle of the last century, Newton's long-established replenishment theory of comets was considered to have been completely vindicated. Celestial portents had been well and truly tamed!

The Benevolent Cosmos—Humans Reign Supreme

The confidence of nineteenth-century physicists is legendary. To comprehend this, one has to appreciate the atmosphere of commitment and faith that had grown up around Newtonian science. Thus, within about a hundred years, a situation had emerged in which remarkably simple mathematical laws (those of motion and gravity) were appearing to provide absolute control over the forces of nature. Such a situation was virtually unprecedented and created an approach to (Newtonian) science that was hardly distinguishable from religious zeal. Admittedly, if one were to examine the details, there were some difficulties with the mechanical ether and the explanation of electromagnetic phenomena, but the solution of these difficulties was widely assumed to be just a matter of patience and time. The mechanisms involved were, it was thought, likely to be simple and would fall easily into place once there was a proper grasp of all the relevant experimental phenomena. With our knowledge of the surprises yet in store and of the great upheavals in physics that were to come, it is difficult perhaps to grasp this very real feeling of mastery over the forces of nature that prevailed in the nineteenth century. This feeling of mastery also extended to the astronomical environment. Here it was obviously Newton's "other law" and the presumed harmless character of that environment that gave rise to the feeling. Thus, while gravity could be seen to pervade the universe and to be providing a smooth undercurrent of physical control over the behavior of matter, it was also possible so far as comets were concerned to draw a clear demarcation between the unharmed earth on the one hand and the noninterventionist cosmos on the other. The impact of this revitalized Aristotelian arrangement was considerable, for never before had there been created such an impression of security in relation to the astronomical surroundings. The new benevolent cosmos underwritten by Newtonian laws could even be seen as the ultimate triumph of a tranquil Christian tradition over fiercer Judaic traditions and those of presumed less enlightened faiths. It was no accident, of course, that an Anglo-Saxon-Protestant-dominated empire at its zenith should have provided the setting within which the universe was to be comprehended. With everything under such control, it is no surprise that earth scientists and biologists now felt free to explain terrestrial evolution and biological evolution without the hindrance of any thoughts of external interference. In a like manner historians and social scientists modified their view of historical evolution and looked more and more to human factors dominating the course of events.

There is a widespread and mistaken view nowadays, especially amongst earth scientists and biblical fundamentalists, that the issue of catastrophism in earth history was decided at this epoch (i.e., the middle of the nineteenth century) by geologists and biologists debating the forces of natural selection. In actual fact, the uniformitarian framework had already been agreed upon by astronomers, as we have seen, and it had simply become a matter of convincing oneself that contemporary terrestrial processes were enough to explain evolutionary change, as revealed for example by the fossil record. Admittedly, for a while, French catastrophists like Laplace and Cuvier opposed the trend, but a short-lived Napoleonic empire was hardly a firm enough base for countering the uniformitarian advance. As it happens, the geologist Lyell and the natural historian Darwin and their followers, took the lead over this issue because they were not particularly bound by any biblical or physical conventions regarding the age of the earth; but insofar as they needed support for the slow action of virtually undetectable forces, the physical scene had already been set.

Likewise, it is hardly accidental that human history was now presumed by experts to be largely due to the action of unseen social forces acting either progressively, through a consensus of similarly oppressed minds, or randomly, through leads from solitary disturbed minds. Indeed, with the problems of physical science apparently on the verge of resolution, the social sciences seemed now to acquire a new attraction. A new cultural division arose in which the mastery over physical science was taken for granted, and a justification apparently existed for the exclusive study of the humanities. Leaders of the new thinking such as Marx and Freud naturally gravitated to the country of origin of the new paradigm, and so on: the point to be made here is that none of these developments molding the twentieth century outlook on the nature of the world could have arisen without the vital sense of control over natural phenomena that had been acquired through physical and astronomical science. Celestial portents had been more than just tamed, they had become unthinkable.

Although astronomical facts lay at the root of this formidable new universe, the apparent control over the forces of nature that came to be so widely accepted eventually gave way to a somewhat dismissive attitude on the part of physicists towards some aspects of astronomy. Unlike earthbound experiments where some degree of regulation is possible, the complexities of the astronomical scene have to be accepted more or less at face value, and one can detect a growing impatience with the more speculative style of inquiry that is then necessary. As a result, the older field of physics and the sibling discipline of astrophysics tended to move apart. Coexistence was always possible, however, so long as astrophysicists confined their activities to the astronomical scene and physicists and others confined their activities to the terrestrial scene. In practice this is exactly what happened: physicists concentrated on the structure of fundamental particles and the properties of matter that could be studied in the laboratory, while astrophysicists concentrated on the properties of stars and larger systems like galaxies and the universe as a whole. Only in recent years have leading practitioners in these fields sealed this happy arrangement by developing a mutual interest in the common forces operating in the deepest recesses of nuclear structure and in the most inaccessible phase of the universe's supposed evolution—the first few seconds of existence following the big bang.

While it is a source of strength to these practitioners that their inevitably unconstrained theorizing in these fields has little or no impact on other areas of knowledge, it is quite remarkable that such developments should have taken place with little or no regard for the more obvious and more accessible area of overlap between physics and astrophysics, namely the planetary and interplanetary region where comets make their appearance. The fact is, however, that for at least half the present century and despite the great advances in physics since then, the planetary system and its environment were regarded as virtually inert and of little fundamental significance to science. Indeed, this was so much the situation midway through the twentieth century that when the opportunity arose to pursue planetary and interplanetary investigations with space probes, many physicists and astrophysicists were quite ready to agree that "space was bilge." It was as if Newton's "other law" had proved itself such a perfect recipe for maintaining order that there was a general reluctance to uncover anything that might disturb the status quo.

That this attitude undoubtedly prevailed is illustrated rather well by the scientific response at this time to an unheralded attempt by Velikovsky to reinstate the ancient fear of comets. This author in effect sought to revive the discussion of catastrophism that took place during the period of Enlightenment, but he was not able to identify the irrational component of Newton's argument (i.e., the replenishment theory) and laid himself open to easy scientific criticism by choosing to doubt various aspects of gravitational theory. During the ensuing vitrioloic exchanges of opinion however, it also became apparent that many modern scientists were defending a fundamentally innocuous role for comets without an entirely clear reason for so doing. Thus, although there was evidence that comets could produce meteor streams, the progenitors in many cases had not been detected; and it was not beyond the bound of possibility that invisible remnants of a more destructive kind were also produced by comets. It was indeed obvious that a centuries-old assumption was without a secure observational basis and that there was no absolute guarantee that Newton's "other law" is correct.

A Dilemma Resolved

By this time, of course, the space age had arrived and the importance of past impact cratering on planetary surfaces had begun to be recognized. The extent of the asteroidal population in earth-crossing orbits was also being revealed. It was not long, therefore, before these new discoveries were thought to imply a cause-and-effect relationship. Previously unobserved bodies greater than a kilometer in size were indeed impacting on the earth every million years or so, on average. This was certainly more frequently than observed comets, but still seemed to be of such little consequence on the time scale of historical events that it was considered rather unlikely to have any bearing on the subject of catastrophism. The new findings tended therefore to excite little attention and would perhaps have continued in this vein had not the source of earth-crossing asteroids proved rather problematic. Thus, from the time of their discovery, it was assumed that Jupiter must be responsible for deflecting these earth-crossing asteroids from supposed original orbits in the asteroid belt. The assumption has persisted until the present despite a continuing failure to demonstrate how the mechanism works. Under the circumstances, one might have expected the alternative possibility, that is, that earth-crossing asteroids are "dead comets," to be welcomed; but the belief that all comets turned into dust was evidently still too strongly imprinted to allow such a change of view. Not that the resistance to this change was necessarily irrational; for by this time, the presumed physical nature of comets had been calculated so as to conform with other developments in modern astrophysical theory. Accordingly, all observed comets were now assumed to have been originally produced in the primordial solar nebula where the formative interstellar medium is thought to have been sufficiently dense to allow small particles of dust to accrete into larger bodies while also absorbing a variety of more volatile chemicals. Such a process required comets to be very cold throughout their lifetime, and there was in principle no qualitative difference between ordinary kilometer-sized comets and the occasional giant comets with diameters greater than ten kilometers. Thus, it was expected that the meteor streams produced by ordinary comets and giant comets would be qualitatively similar, as required for the presumed uniformitarian astronomical environment of the earth. It followed that there was no good reason for supposing that giant comets might produce asteroids.

According to present-day astrophysics therefore, comets are supposed to form in a primordial solar nebula under conditions of *cold* gravitational collapse. However, this theory is not necessarily supported by all the observations; it is possible, alternatively, that comet-like planetesimals condense out of a hot primordial system that is very much more compressed.[3] Under these circumstances, giant comets in particular are highly differentiated chemically and are qualitatively very different from the smaller more ordinary comets. Giant comets indeed then seem to be very like the primitive bodies out of which observed meteorites are supposed to have formed; but of even greater significance is the fact that contemporary examples (e.g., Chiron) may produce meteor streams that develop by progressive fragmentation into asteroidal bodies of smaller and smaller size. Such meteor streams are potentially much more dangerous than conventional meteor streams since they are fed by a huge swarm of small asteroids of the "Tunguska" type, which may encounter the earth from time to time, producing violent fireball storms and a battery of great explosions in the hundred- or thousand-megaton class—nature's equivalent of a devastating nuclear war! The frequency of such catastrophic encounters depends both on the rate of arrival of giant comets in earth-crossing orbits and on orbital coincidences between an evolving giant comet and the earth. The former is largely controlled by Galactic interactions, and the various predicted periodicities are shown in Table 14-1.[4]

TABLE 14–1.
PREDICTED PERIODICITIES IN THE TERRESTRIAL RECORD

Typical period*	Physical origin
100 ± yr	Giant comet orbital commensurability with earth.
2000 ± yr	Giant comet orbital precession.
30,000 ± yr	Lifetime of disintegrating giant comet.
100,000 ± yr	Recurrence time for giant comets during showers.
10 ± Myr	Stochastically distributed intervals associated with disturbances of the sun's comet cloud by molecular clouds in the galactic plane producing showers.
30 Myr	Vertical oscillations of the sun's orbit through the galactic plane.
250 Myr	Radial oscillations in the galactic plane
2000 ± Myr	Beat frequencies associated with recurrent production of molecular cloud systems

* ± implies approximate order-of-magnitude range.

According to our current knowledge, it is very likely that many of these encounters *are* impressed upon the terrestrial record; so the earth's evolution by implication is catastrophic. However the physical mechanisms involved are not yet fully worked out and some aspects of this picture are necessarily uncertain at the present time. The obliteration of dinosaurs 65 million years ago by an asteroid epitomizes a view of this kind of evolution which has recently caught the popular imagination. Nevertheless the broad picture is reasonably clear: successive giant comets arriving in earth-crossing orbits will each produce a huge meteor stream that for an interval of ten thousand years or so is responsible for periodically maintaining a dense stratospheric dust veil and inducing an ice age; during the subsequent interglacial, the giant comet declines to near invisibility while experiencing disintegrations that give rise to periodic terrestrial bombardments mostly by bodies in the Tunguska and super-Tunguska class.

The question that now confronts us is whether Newton's "other law" is correct, or whether there are indications in the earth's recent history and its current environment that the more violent scenario really applies. Such indications, if present, would tell us that modern astrophysical theory, through its predilection for the noninterventionist cosmos, has succeeded in developing a wholly false view of the origin and importance of comets, predicated on our current understanding of the nature of the universe. The extent to which it would be necessary to preserve this understanding of the universe would then become an open question.

The Most Recent Giant Comet

The largest if not the brightest meteor stream in the sky is the so-called Taurid-Arietid stream. It approaches the earth by night from the (celestial) West in the months of November/December and by day from the North during May/June. Since the earth's crossing time for most other meteor streams is just a few hours, there is a considerable contrast between the breadth of the Taurid-Arietid stream and that of other streams: volume for volume and mass for mass, the source of the Taurid-Arietid stream may not be far short of a million times larger than that of a typical stream. And since a typical comet is a few kilometers in diameter, the ultimate source in this case may easily be a giant comet a few hundred kilometers in size. The material in this stream mostly circulates in elliptical orbits within the path of Jupiter with periods of around three years. Most other meteor streams have longer periods than this, but a substantial fraction tend to concentrate in the same part of the sky as the Taurid-Arietid stream, thereby raising the possibility of a common origin (i.e., fragmentation of a single parent body) and a subsequent more rapid dispersal of those streams that cross the orbit of Jupiter.

It has recently been noted that the Taurid-Arietid stream coincides with an even broader stream of smaller particles that are similar in character to those of the zodiacal cloud; it is likely therefore that this stream is ultimately the major current source of dust in the terrestrial environment. Comet Encke (period: 3.3 years) is a prominent member of the Taurid-Arietid stream; and a significant proportion of the known population of earth-crossing asteroids (around 10 percent) seems also to be identified with the stream. Indeed one of these asteroids is somewhat cometary in appearance. There is a growing impression that the stream derives from a cometary progenitor, and that it must then progressively disintegrate into a variety of asteroidal debris. The Tunguska body of 100,000 tons is a case in point: it struck the earth on the morning of June 30, 1908, and was almost certainly a member of the stream. It is also established now that the stream includes many somewhat smaller bodies with masses in excess of a ton, for these have been detected striking the moon following the placing there of several seismometers by the Apollo astronauts. Such missiles appear to have spread throughout the stream during approximately the last twenty thousand years from a dense swarm at its core. A cloud of escapees from this swarm was in fact encountered during a week-long penetration at the end of June 1975.

To sum up, the wide range of material now observed in the Taurid-Arietid stream is consistent with a giant comet for its source and a history of successive fragmentations periodically replenishing the dense core of the stream and enhancing the zodiacal cloud. Such a history was in fact conjectured for the stream over thirty years ago when astronomers Whipple and Hamid accurately retrocalculated the orbits of a number of meteors to indicate that several major fragmentations had taken place during the last five thousand years. The most significant of these events took place around 3000 B.C. due to an encounter in the asteroid belt; another was deemed to have occurred around A.D. 500, with possibly yet another in the second half of the second millennium B.C. The epochs around 3000 and 1300 B.C. in particular correspond to significant deteriorations in the global climate for two or three centuries or more—at least to the extent that global climate is reflected by the advance and retreat of the northern limit of forestation in Canada and Europe and by the rise and fall of the sea level around England.

It is known from other studies that a correlation exists between global rainfall and the incidence of meteor dust on the earth; so the indications now are for a considerable degree of climatic control by the Taurid-Arietid stream. Should there be confirmation of the strong concentrations of cosmic dust that have been detected in ice core deposits corresponding to the last major glaciation (around 20,000–10,000 B.C.), and of their primitive meteoritic composition that seems to be identical to that of the Tunguska body, then powerful evidence is available of the way in which a particular giant comet, namely the most recent, has produced the last ice age and continued to modulate the climate during the subsequent interglacial through the intermediary of stratospheric dust veils.

In essence,then, we are learning from a considerable variety of indicators that a strong correlation exists between current properties of the astronomical environment and the earth's history during the last twenty thousand years: a giant comet has been the dominant controlling influence over the earth's evolution during this period. We may note in passing that the most recent mass extinction of biota (mammoths, etc.) occurred only twelve thousand years ago during the last glaciation, apparently while a temporary climatic amelioration was in progress. It is a process that may therefore represent in microcosm what has been going on continuously, albeit with the modulations tabulated above, throughout the earth's history.

The provenance of ice ages, particularly the most recent, has been a matter of contention among geologists for many years. Long before the Milankovitch hypothesis was developed (with its system of delicate feedbacks and hysteresis effects), there was a preference for dust particles in space determining the course of climatic change. No suspicion of the role of giant comets had appeared at this time but a prescient geologist has recently remarked that "if a dust chronology were available, scientists could check it against the ice-age chronology." This is certainly a pertinent comment, for a time has now arrived where the wholesale retrocalculation of meteor orbits in the Taurid-Arietid stream is possible; it may shortly enable us to identify the most significant fragmentations during the Holocene and correlate them with climatic variations and other terrestrial events. With the latter particularly in mind, we would also be in a position to examine the human record.

The Human Record

Thus, before 3000 B.C., our ancestors would have regularly observed at least one large comet in the sky. It was probably a brilliant though essentially harmless spectacle, but also frequently an awesome one when orbital coincidences brought the earth particularly close. Indeed, its most conspicuous characteristic would have been the regularity of its motion, the commensurability of its orbit with that of the earth. It would be expected that the earliest references to this comet would recognize its capacity as a timepiece (cf. Chronos). With a history of progressive splitting, one might expect several subsidiary giants in similar orbits at this time, surviving for centuries or even a millennium. Such a family of gods in the sky, their eternal stream, and the earth's predictable and splendid encounters, would inevitably generate a lasting sense of union between heaven and earth. At the epoch in question, however, around 3000 B.C., a major fragmentation of the primary body would produce an additional battery of comets; it would not be surprising if onlookers subsequently thought they were witnessing a battle for mastery over the sky, and that this was in some way associated with the assaults on the earth that inevitably followed. These assaults due to encounters with the core of the stream would in effect be global bombardments by Tunguska and super-Tunguska type bodies that would leave an indelible memory for the surviving humans and a lasting fear of the gods in the sky.

Such memories would be of local floods and widespread destruction by fire over areas the size of a nation. More or less contemporary migrations would ensue all over the world, with some land lying sterile for centuries. There would be a trend towards defensive building both against assaults from the sky and against opportunist attacks from human marauders. In time, with all but a few vestiges of the violent destruction erased, there would only be an extended period of extreme social disorder—the end of a golden age, or a dark age—to mark the course of history and the impact from above.

We can well understand how a frenzied response might arise in the form of religious temples to propitiate the violent gods and astronomical observatories to anticipate future returns, though we have until now always assumed it was mere calendric or navigational requirements that arbitrarily inspired the growth of astronomy at these times, and mere technology that inspired a new generation to produce the pyramids and Stonehenge. With the passage of time, of course, the encounters would weaken and there would be only fireballs and declining comets to remind one of the former events. But at some stage there might at any time be a further fragmentation and a revival of the earlier terror. The scribes would scan the ancient records and attempt to prophesy the course of events, but eventually there would be no avoiding further multiple bombardment by Tunguskas and super-Tunguskas, with massive destruction of cities and widespread incineration of crops and land. Massive migrations would again take place as survivors seek to escape, and doubts would be raised concerning the efficacy of prayer; a dark age would follow and a renaissance in due course with new questions about religion and cosmology. And then, with the passage of time, the whole process can be expected to repeat itself, continuing until the core of the giant comet has been completely whittled away. We might even envisage destruction so great, and dark ages so effective, that only the dimmest memories will later exist of giants that once walked the earth, of heavenly clouds that a creator once built in the sky, of prophets and messiahs who warned of doom and salvation, and of floods and cataracts of fire that were used to cleanse the earth. We might anticipate that the intellectual confusion would reach new heights when the comet-asteroid deities known to be responsible for all the mayhem finally disappear from sight—which they apparently did during the first millennium B.C. We might anticipate the worship of new invisible gods or the diversion to new planetary gods. We might even learn to agree with an Aristotle, a Ptolemy, or a Newton, as they seek to dismiss the thunderbolts of a previous generation and restore a sense of order in heaven and on earth.

The peak of the observed fireball flux, superior to the present day's, coincided with the Taurid-Arietid stream, according to European records, a century or two after the time of Emperor Charlemagne and King Alfred. Indeed, while the latter fought to bring their peoples out of a deep dark age, Chinese rulers of the T'ang Dynasty held astronomers under tight control, their duties being to prognosticate the future on the basis of signs in the sky such as meteors and comets. It was a period during which a remarkable wealth of brilliant fireballs was recorded. On one occasion, for example, there was noted "a great shower of meteors of all sizes, lasting through the night," and on another soon after "dozens of small stars crisscrossing the sky through the night." At the same era, on the other side of the earth, the Anglo-Saxon Chronicle was being recorded and for the month of June A.D. 793, it was noted that "fierce, foreboding omens came over the land of Northumbria and wretchedly terrified the people. There were excessive whirlwinds, lightning storms and fiery dragons were seen flying in the sky. These signs were followed by great famine. . . ." The very meagerness of the information available to us now may be a testimony to the destructive forces at play.

During the classical era too, we fail to understand the earliest natural philosophers and poets if we do not recognize their concern for the weapons and missiles brandished by the gods. Meteorology then was not the wholly earthbound science that it has now become but the domain of astronomical agents who were only brought to the ground by Aristotle. Again we have many explicit statements by Chinese astronomers who warn us of the nature of the contemporary sky:

> Dynasty Han, Reign Yuan-yan, Year 1, Month 4, Day Ding-you (i.e., 22 May, 12 B.C.). At the hour of rifu (i.e., 3–5 P.M.), the sky was cloudless. There was a rumbling like thunder. A meteor with a head as big as a *fou* (an earthenware pot), and a length of some ten-odd *zhang* (a *zhang* is 12 degrees), color bright red and white, went southeastward from below the sun. In all directions, meteors, some as large as basins, others as large as hens' eggs, brilliantly rained down. This only ceased at evening twilight.

We cannot yet predict the future; the core body of the Taurid-Arietid stream has still to be observed. Indeed, the big guns of modern science are not trained in this direction at all. But there is a swarm of boulders out there and a future confrontation with a barrage of Tunguskas is a very reasonable projection from the state of current knowledge. At least one form of star wars can be virtually guaranteed, and only time will tell whether we face a nuclear winter or a cometary winter.

NOTES

1. Useful introductions to the material in this section may be obtained from B. Farrington, *Greek Science*, Pelican, 1949; O. Neugebauer, *History of Ancient Astronomy: Problems and Methods*, 1945; B. L. van der Waerden, *Science Awakening*, Oxford, 1974; and F. Cornford, *Principium Sapientiae*, Peter Smith, 1971. Other references appear in M. Bailey, S. Clube, and W. Napier, "The Origin of Comets," in *Vistas in Astronomy*, Pergamon, 1986. For an indication of continuity between the myths of ancient Near Eastern people and Greek ideas, see the R. Patai paper in this book.

2. For background material for this section, see P. Hazard, *The European Mind 1680–1715*, Hollis and Carder, 1953; and L. Stecchini, "The Newton Affair," *Kronos* IX, no. 2 (1984): 34ff.

3. The suspicion that a conceptual error in physics led to a misunderstanding of the true nature of galaxies and hence of the origin and evolution of the Solar System was raised in S. Clube, "Does Our Galaxy Have a Violent History?" in *Vistas in Astronomy* (Pergamon, 1978), 22, 77; and "The Material Vacuum," *Monthly Notices of the Royal Astronomical Society* 193 (1980): 385.

4. The implicit new setting for terrestrial history in which giant comets prove to be the dominant evolutionary force has recently been reviewed in "Giant Comets and the Galaxy: Implications of the Terrestrial Record," in *The Galaxy and the Solar System*, Tucson: The University of Arizona Press, 1985. A more popular exposition of this view of terrestrial catastrophism was presented by these authors in *The Cosmic Serpent*, New York: Universe, 1982, and London: Faber, 1982, in which they indicate some of the possible implications for human history of the last five thousand years.

15

Comets and Cataclysmic Changes

WILLIAM H. STIEBING, JR.

Professor Clube contends that giant comets split apart to create meteor streams whose large bodies bombard the earth periodically, destroying species and producing climatic changes. The most recent cometary splitting and bombardment has occurred within the last five thousand years and is reflected in myths and human beliefs in heavenly portents. While we cannot yet determine the validity of Professor Clube's theory of cometary origins, it seems likely that major impacts on the earth have punctuated geological history at intervals of millions of years. But there is little evidence to support the view that such cosmic catastrophes have taken place since c. 3000 B.C. Ancient Mesopotamian, Egyptian, and Chinese references to astronomical phenomena give no hint of major cataclysms or fear of comets. There is also no archaeological evidence of cosmic collisions that might have given rise to early myths about warring gods or universal floods.

Professor S. V. M. Clube introduces his thesis by providing a brief history of the development of cometary theory. He contends that the demise of catastrophism in earth history was primarily due to acceptance of Newton's "other law," which held that comets were not only harmless, but also beneficial to the Solar System. However, Clube's analysis overestimates the role astronomical uniformitarianism played in winning acceptance of geological and biological uniformitarianism. The astronomical view of an orderly universe did influence the thought of early earth scientists and biologists, but Professor Clube neglects other currents of thought and other discoveries that occurred in the seventeenth to nineteenth centuries. For example, in 1669 (*before* Newton proposed his "other law") Nicholas

Steno's observation that most fossil deposits contained creatures that normally inhabited the same environment led him to conclude that geological strata and their fossil contents had been deposited by a natural process, not by catastrophes such as great floods. His contemporary, Robert Hooke (originator of a wave theory of light and a kinetic theory of gases, among other things), agreed that fossils were natural deposits that could be used to reconstruct a chronology for the earth much longer than that suggested by the Bible. These ideas influenced James Hutton's uniformitarianism (1785) and that of Lyell (1830–33), as did the discovery that prehistoric man was contemporary with some extinct species. The antiquity of man was denied by eighteenth and early nineteenth century catastrophism, which was intent on upholding the biblical chronology for man's creation. Professor Clube's historical sketch pays too much attention to Newton and not enough to the interplay of science, early archaeology, and religion. Fortunately, this does not seriously affect his theory.

There are two major theses in Professor Clube's very interesting paper. The first is that there are giant comets that are qualitatively different from the smaller, more ordinary comets. Through fragmentation these giant comets may produce meteor streams whose asteroids probably encountered the earth a number of times in the past, extinguishing many species and inducing ice ages. The second contention is that the most recent giant comet to approach the earth produced the Taurid-Arietid meteor stream. This giant comet split a number of times between c. 3000 B.C. and 500 B.C., with especially severe episodes around 3000 B.C. and 1300 B.C. The asteroids produced by the breakup of this comet periodically rained down upon the earth, destroying cities, forcing migrations, and producing dark ages.

Retrocalculation of meteor orbits provides persuasive evidence that several fragmentations have taken place within the Taurid-Arietid meteor stream during the past five thousand years, and that these fragmentations correspond in time with periods of climatic deterioration on earth, especially around 3000 B.C. and 1300–1150 B.C. Eventually it may be possible to show by retrocalculation that a fragmented giant comet is the source of the entire Taurid-Arietid stream. But at present Professor Clube's description of the formation of giant comets and their fragmentation into meteor streams is only speculative, as are all other theories of cometary origins. Astronomers and earth scientists, not an archaeologist/ancient historian like the present writer, eventually will determine whether or not his hypothesis is valid.

While the basic portions of Professor Clube's theory cannot yet be demonstrated, what of the claim that the earth has experienced major cataclysms of cosmic origin a number of times within the past five thousand years? Here the evidence does not seem to support the catastrophist scenario.

Professor Clube suggests that the appearance of temples, astronomy, observatories, the pyramids, and Stonehenge soon after 3000 B.C. was due to the spectacle of comets splitting and assaulting the earth at that time. If all these features were due to comet-inspired terror, why do comets play such a minor role in the religions of the earliest civilizations? The gods and goddesses of the ancient world were often associated with the major heavenly bodies—the sun, moon, stars, and planets. But none of the ancient cultures seems to have had a comet god as a major figure in its pantheon! Velikovsky avoided this problem by having his comet become the planet Venus, but the failure of the ancients to worship comets or shooting stars is difficult to explain under Clube's hypothesis.

What evidence do we have that around 3000 B.C. strange phenomena in the skies caused people to begin paying special attention to the movements of heavenly bodies and a few generations later to construct monuments such as the pyramids and Stonehenge? None at all. Buildings that were clearly temples were in use in Mesopotamia by at least the Ubaid Period (c. 4500–3500 B.C.).[1] Temples continued to develop and increase in number as new settlements and cities came into existence, but there was no sudden or unusual burst of temple building around 3000 B.C. From the Ubaid Period onward, Mesopotamian temples were built atop temple platforms called ziggurats, which gradually grew higher and higher. Pyramid-like ziggurats (which sometimes seem to have served in later times as astronomical observatories) did not appear in their fully developed form until c. 2100 B.C.[2] On the other hand, the pyramids of Egypt were *tombs* designed to ensure the immortality of pharaohs. Their development, beginning c. 2650 B.C., is gradual and seems to have nothing to do with a new awareness of catastrophes caused by heavenly bodies.[3] And while Stonehenge may have developed as an astronomical observatory between c. 2500 B.C. and 2000 B.C., its structural elements were used to record the *regular* (non-catastrophic) movements of the sun, moon, and stars.[4] Pyramids and astronomical observatories in the New World developed even later (c. 1200 B.C.).[5] There is simply no evidence to support the view that around 3000 B.C. cosmic catastrophes caused the sudden appearance of temples, observatories, or other monuments.

Nor did astronomy appear soon after 3000 B.C. The Sumerians (who between 3500 B.C. and 3000 B.C. created the earliest civilization known at present) seem to have had little interest in astronomy. All they have left us is a list of about twenty-five stars.[6] The earliest real evidence we have for Mesopotamian astronomy comes from the Old Babylonian Period (c. 1800–1600 B.C.). Texts from this era include the famous observations of the appearances and disappearances of Venus made during the reign of Ammisaduqa, and tablets grouping constellations, stars, and planets into three zones of the sky, each zone containing twelve sectors. From the Kassite and Assyrian Periods (c. 1600–612 B.C.) we have a collection of omen texts known as *Enuma Anu Enlil*, a number of "Astrolabes" listing stars in their sectors, and the star and planet catalogue known as *Mulapin*. During the Neo-Babylonian and Persian Periods (612–331 B.C.) mathematical astronomy as well as zodiacal and horoscopic astrology were developed,[7] providing the base for the development of Hellenistic astronomy. This gradual growth of astronomy and astrology over some two thousand years seems to owe nothing to terror produced by cometary bombardments.

The Egyptians may have been observing the rising of Sirius as early as 3000 B.C., for a First Dynasty tomb contained a plaque describing Sirius as "herald of the new year and of the flood."[8] It is not known when the Egyptians first recognized the relationship between the first visibility of Sirius in the morning sky and the beginning of the Nile flood. But for most of their history they do not seem to have used their observations of the heavens for much except calendrical functions and determining cardinal points of the compass for orienting buildings. The skies were generally not the source of fearful portents for the Egyptians. There are no signs that they accepted astrology (and the careful astronomical observations it implies) before the Hellenistic Period when Egyptians, like the Greeks, came under the influence of Mesopotamian astronomy.[9]

The Chinese also produced early astronomical records, but only from the fourteenth to thirteenth centuries B.C. onward. The oldest record of a nova occurs on a Chinese oracle bone from c. 1300 B.C., and the stars, constellations, and lunar phenomena are also mentioned in these early texts.[10] One inscription dated 1211 B.C. states that "the summer solstice recurs at 548 days after the winter solstice," indicating that regular and long-continuing observations were certainly being carried out by the thirteenth century B.C.[11] However, the earliest Chinese records of comets and meteor showers date only from the seventh century B.C.[12] This fact does not seem to indicate that the early Chinese became interested in heavenly bodies because of the activities of giant comets and the asteroid bombardments they caused.

Clube's hypothesis would lead us to expect numerous references to unusual phenomena, to splitting comets and deadly showers of asteroids, especially in the Mesopotamian and Chinese omen or oracle texts. But such references do not occur. In Mesopotamian omen texts the appearance of clouds and other meterological phenomena are viewed as portents almost as often as astronomical phenomena are,[13] and the astronomical phenomena that are mentioned are generally quite ordinary. Comets are conspicuous by the relative infrequency of references to them. It should also be noted that shooting stars were either good or bad omens for Mesopotamians, depending on the area of the heavens where they were seen.[14] (The ancient Romans also regarded comets as good omens at times.)[15]

The earliest Chinese texts likewise show no particular interest in comets and meteors. Even when Chinese astronomers started recording these phenomena, their descriptions are objective and scientific, not terror-filled. The Chinese account of the large meteor shower on May 22, 12 B.C. quoted by Professor Clube in his paper is a case in point. Another is a description from 1064 A.D. of a meteorite whose landing produced an oval crater and started some small fires.[16] The phenomena are carefully described, but there is no indication that past experience had led the astronomers to expect that chaos, widespread catastrophes, or an end to civilization would follow. The facts do not seem to support Professor Clube's contention that the ancients were led to scan the heavens because of their fear of comet-related catastrophes.

In his paper Professor Clube negates this lack of ancient references to cosmic cataclysms by suggesting that the cometary destructions would have produced dark ages—in some cases so dark that only the dimmest memories of the events survived in myths and legends. This is essentially the same answer Velikovsky provided for this problem (he called the phenomenon "collective amnesia"). Creation myths, flood stories, and stories of conflicts between gods and goddesses must be used to supply the evidence for the thesis.

Use of such ambiguous evidence was similarly defended in the popular treatment of the giant comet thesis that Clube wrote with Bill Napier. They comment on the paucity of Mesopotamian and Egyptian references to comets: "This cannot be because they did not exist, so it must be because they were generally described as something else."[17] Not only does this statement exhibit faulty logic (the conclusion is *not* the only one that can be drawn from the lack of ancient Egyptian and Mesopotamian references to comets), but it also has the effect of making the hypothesis of historical cometary catastrophes unfalsifiable. If there is evidence of catastrophes it can be used to support the hypothesis. But if there is *no* clear evidence of cometary catastrophes, then that also proves the theory, since it shows that people were so terrified of comets that they could not refer to them openly or that the dark age created by the catastrophe was so deep that only dim memories of the event survived. How does one argue against such a view? Its "logic" reminds one of the story about a citizen of a large American city who always wore an amulet to keep wild elephants away. When his friends, trying to reason with him, pointed out that there were no wild elephants on the entire continent, the man was unfazed. "Of course there aren't," he responded. "See how well it works!"

Moreover, not only does the ancient textual evidence provide little support for a theory of historical cosmic disasters, but archaeological evidence is against such a theory. If Mesopotamian, biblical, and other flood myths derive from cosmic catastrophes that occurred around 3000 B.C., then there must have been extensive flooding of Near Eastern areas at that time. But such floods are not evidenced in the archaeological record. When Leonard Woolley uncovered a thick flood deposit during his excavations at Ur in southern Mesopotamia in the 1920s it was interpreted as evidence of a great flood that had covered most of the Tigris-Euphrates Valley in early times. Soon afterward flood deposits were discovered at other Mesopotamian sites and a pan-Mesopotamian flood seemed to be proved. Clube and Napier claimed this great flood as evidence of the catastrophes caused by the breakup of a giant comet c. 3000 B.C.[18] But further study has shown that the various prehistoric flood deposits of Mesopotamia belong to different ages, most of them earlier than 3000 B.C. And some Mesopotamian sites, including al-'Ubaid near Ur, exhibit no flood deposits at all.[19] The flood layers that were found are the results of several local floods—there was no single prehistoric flood that covered most of Mesopotamia. There is also no evidence of a great flood at other Near Eastern sites (Jericho, for example) occupied this early. There does not seem to have been a single great flood that gave rise to biblical, Mesopotamian, and other early flood stories.[20] There is also no widespread destruction of cities by fire or other causes about this time. Rather, the era just after 3000 B.C. was a period of intensive cultural growth and development in the Near East as both the Sumerian and Egyptian civilizations entered into their early dynastic periods.

Professor Clube's astronomical theories are intriguing and, if valid, may help explain some of the mass extinctions of species in past geological ages. More limited contacts between the earth and meteor streams containing large amounts of space dust might also explain less radical climatic changes such as the ones that seem to have taken place c. 3000 B.C. and c. 1300–1150 B.C. or the so-called "Little Ice Age," which began c. A.D. 1550 and ended only recently (c. 1915–20).[21] But he has extended his theory beyond the limits of the evidence in arguing for *major* cataclysms during the past five thousand years. Archaeological and historical sources do not support his scenario, and subjective interpretations of myths and legends simply cannot substitute for hard evidence.

NOTES

1. G. Roux, *Ancient Iraq* (New York: Penguin Books, Second Edition, 1980), 70.

2. W. Stiebing, Jr., *Ancient Astronauts, Cosmic Collisions and Other Popular Theories About Man's Past* (Buffalo, New York: Prometheus Books, 1984), 121–122.

3. *Ibid.*, 110–116.

4. Stonehenge's primary astronomical orientation is to the rising of the sun at the summer solstice. It may have been used as a gigantic computer to determine matters of calendrical importance. See C. Renfrew, *Before Civilization: The Radiocarbon Revolution and Prehistoric Europe* (Cambridge: Cambridge University Press, 1979), 214–247.

5. W. Stiebing, Jr., *Ancient Astronauts, Cosmic Collisions*, 123–124.

6. S. Kramer, *The Sumerians* (Chicago: University of Chicago Press, 1963), 90–91.

7. O. Neugebauer, *The Exact Sciences in Antiquity* (Providence, Rhode Island: Brown University Press, Second Edition, 1957), 99–144; and B. van der Waerden, *Science Awakening, II: The Birth of Astronomy* (Leiden: Noordhoff International Publishing, 1974), 46–126.

8. B. van der Waerden, *Science Awakening*, 8.

9. O. Neugebauer, *The Exact Sciences in Antiquity*, 80–91.

10. J. Needham, *Science and Civilization in China. Volume 3: Mathematics and the Sciences of the Heavens and the Earth* (Cambridge: Cambridge University Press, 1959), 242–245, 424.

11. *Ibid.*, 293.

12. *Ibid.*, 430–434.

13. R. Culver and P. Ianna, *The Gemini Syndrome: A Scientific Evaluation of Astrology* (Buffalo, New York: Prometheus Books, Revised Edition, 1984), 15–18.

14. A. Oppenheim, *Ancient Mesopotamia* (Chicago: University of Chicago Press, 1964) 219.

15. For a Roman interpretation of a comet as a good omen see M. Grant, *The Twelve Caesars* (New York: Charles Scribner's Sons, 1975), 55.

16. J. Needham, *Science and Civilization in China*, 433–434.

17. V. Clube and B. Napier, *The Cosmic Serpent* (New York: Universe Books, 1982) 163.

18. *Ibid.*, 209.

19. J. Bright, "Has Archaeology Found Evidence of the Flood?" *Biblical Archaeologist*, 5, no. 4 (Dec. 1942): 55–62.

20. W. Stiebing, Jr., *Ancient Astronauts, Cosmic Collisions*, 9–22.

21. R. Bryson and C. Padoch, "On the Climates of History," edited by R. Rotberg and T. Rabb, in *Climate and History* (Princeton, New Jersey: Princeton University Press, 1981), 3–17.

REFERENCES

Bright, J. "Has Archaeology Found Evidence of the Flood?" *Biblical Archaeologist* 5, no. 4 (Dec. 1942): 55–62.

Bryson, R. and C. Padoch. "On the Climates of History." Pp. 3–17, edited by R. Rotberg and T. Rabb, in *Climate and History*. Princeton: Princeton University Press, 1981.

Clube, V. and B. Napier. *The Cosmic Serpent*. New York: Universe Books, 1982.

Culver, R. and P. Inna. *The Gemini Syndrome: A Scientific Evaluation of Astrology*. Buffalo, New York: Prometheus Books, Rev. ed., 1984.

Grant, M. *The Twelve Caesars*. New York: Charles Scribner's Sons, 1975.

Kramer, S. *The Sumerians*. Chicago: University of Chicago Press, 1963.

Needham, J. *Science and Civilization in China, Vol. 3: Mathematics and the Sciences of the Heavens and the Earth*. Cambridge: Cambridge University Press, 1959.

Neugebauer, O. *The Exact Sciences in Antiquity*. Providence, Rhode Island: Brown University Press, Second edition, 1957.

Oppenheim, A. *Ancient Mesopotamia*. Chicago: University of Chicago Press, 1964.

Renfrew, C. *Before Civilization: The Radiocarbon Revolution and Prehistoric Europe*. Cambridge: Cambridge University Press, 1979.

Roux, G. *Ancient Iraq*. New York: Penguin Books, Second edition, 1980.

Stiebing, Jr., W. *Ancient Astronauts, Cosmic Collisions and Other Popular Theories About Man's Past*. Buffalo, New York: Prometheus Books, 1984.

Van der Waerden, B. *Science Awakening, II: The Birth of Astronomy*. Leiden: Noordhoff International Publishing, 1974.

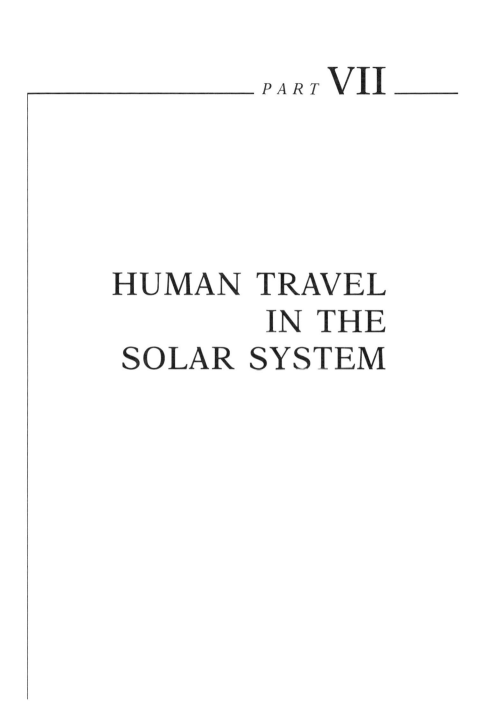

PART VII

HUMAN TRAVEL
IN THE
SOLAR SYSTEM

16

Human Exploration and Development of the Solar System: The First Steps

BRIAN O'LEARY

Mankind is on the threshold of an explosive renaissance in space. Opportunities for humans to explore the Solar System and to use the resources of other celestial bodies will open a multi-trillion-dollar economy, provide a new basis for international cooperation, and create an uplifting of the human spirit.

This paper describes in detail three early targets of opportunity: the moon, the earth-approaching asteroids, and the moons of Mars. It shows how all three could provide both the exploratory and economic basis for the flourishing of culture beyond the earth as soon as the turn of the millennium.

I propose that the most productive next step should be an international human mission to the moons of Mars. Such a mission will open our eyes to the wonders of Mars and provide the economic basis for industrializing space, utilizing easily accessible extraterrestrial materials, free from the gravity bondage of earth.

As these next steps are accomplished, it will become possible to create widespread space tourism; space colonies; growing bases on the moon, Mars, and other bodies; terraformed environments; space agriculture; asteroid mines; and solar power satellites to provide energy to the earth. The forthcoming human migration into space can be likened to the time when our fish ancestors first crawled onto land. The next major step in human evolution is upon us in the new environmental niches of free space and other celestial bodies. The forthcoming space renaissance is a metaphor for human transcendence.

Introduction

Mankind is poised to take its next major step in evolution. As early as the turning of the millennium, some of us may be leaving our home planet permanently. Our ultimate visions of spaceflight will become a reality once we free ourselves from the gravity bondage of the earth and build a growing economy dependent on abundant accessible natural resources of the Solar System. An inevitable by-product will be the hands-on experience of walking on other worlds and the lifting of the human spirit in exotic new settings.

Spinning space settlements up to several kilometers in size will permit thousands to millions of people to create their own environment—climate, lighting, flora, fauna, and for the first time gravity, will be variables from which to choose. Large domed colonies on planets, moons, and asteroids will house the first settler-explorers. Space travel will soon become as easy as air travel, and space tourism may soon become a multi-billion-dollar industry.

These visions are not the idle thoughts of a science fiction fantasy writer, but the rational and obvious projection of practical scientists, engineers, and entrepreneurs. That we are on the threshold of an explosive human renaissance in space is clear. It is no longer a question of whether or not we do it; it is a question of when, how, and by whom it will be done.

While the visions are realistic, carrying them out in a logical sequence seems elusive. Governmental focus on military competition requiring similar technology (but on a far grander scale) and corporate desire for near-term profits have tended to distract us from taking the steps required to fulfill our destiny among the planets. In order to establish a self-sufficient space economy, most of the technology is available and the investments are modest, about 2 percent of the United States federal budget. But we need to have the foresight and vision to make all this happen, and awareness among world leaders has not yet reached that stage.

The purpose of this paper is to examine in detail three early destinations in the Solar System, any one of which would be sufficient to open the way toward a space renaissance. I review studies and concepts and estimate costs, timelines, and pathways toward our ultimate dreams. The three destinations are the moon, the asteroids, and Mars and its moons. I suggest that Phobos and Deimos (hereafter frequently referred to as PhD), the moons of Mars, would be the most exciting, immediate, and cost-effective destinations that would effectively dovetail rationales (resource recovery and exploration) and set the stage for international cooperation.

Initiatives in the United States and Soviet Union

In the 1990s, on the 500th anniversary of Columbus's discovery of the New World, the United States will begin to launch its own new world into space. President Reagan and Congress have given approval to NASA to spend $16 billion on a modular tinkertoy assemblage that will comprise the first permanently-manned space station. The meaning of this decision goes beyond the space station itself. Around the turn of the millennium the space station will become a base camp for assaults to destinations beyond: the moon, Mars, its satellites, the asteroids, the other planets, and eventually the stars.

The space station decision has prompted renewed activity on these longer-term space goals in the United States. In April 1984 a number of leaders in the space community held a closed conference in Los Alamos, New Mexico, to create plans to develop a lunar base as the next major space step for the United States beyond the space station.[1] Out of this conference came a NASA-sponsored technical Symposium on Lunar Bases and Space Activities in the 21st Century, held in Washington, D.C. in October 1984.[2] A second meeting took place in Houston in 1988. Hundreds of scientists and engineers participated.

George Keyworth, then director of the White House Office of Science and Technology Policy, wanted to "recapture the vision of Apollo" and placed a return to the moon high on his list. "The lunar base," said Keyworth, "is one of the more bold and exciting goals that we can reach" in the years ahead.[3]

So much was the White House interested in looking at the lunar base that two months prior to President Reagan's January 1984 State of the Union message, they asked the National Science Foundation to commission two one-month studies that would address the issue of a manned lunar science base as an alterative to space station science.[4] Since then, three more NASA-funded study groups have looked at the lunar base as the next major step in space for the United States after the space station.

The prospect of human voyages to the vicinity of Mars has received even more attention in politically high places, both in the United States and the Soviet Union. In a more recent statement, former U.S. presidential science adviser Keyworth suggested the possibility of putting Mars ahead of the moon. "Now is the time for bold, long-range thinking", he said. "One challenge well within our technological group is a manned Mars mission . . . can we do Mars without the moon first?"[5]

The Soviets appear to be headed for Mars as their high-priority celestial target as evidenced by their official claims, their numerous unmanned probes, their long-duration Salyut rehearsals, and the imminent flight of a gigantic earth-to-orbit booster. Moreover, they were the first to send an unmanned reconnaissance probe to the Martian moon Phobos in 1988. Recent studies have shown that the first human missions to the tiny Martian moons would cost considerably less, would provide more scientific data, and open the way sooner to space industrialization than attempting a Mars landing.[6]

Asteroids have also been proposed as early post-space-station destinations for human missions.[7] Because of their low gravities (as in the case of Phobos and Deimos) and the variety of materials and orbits to choose from, the asteroids could provide abundant resources for the early onset of space industrialization.

One recent study assessed the costs of near-term manned missions to the Martian surface, a near-earth asteroid, and lunar base reconnaissance.[8] The costs were estimated at $39 billion for the Mars missions, $16 billion for the asteroid mission, and $16 billion for the lunar mission. A manned Phobos-Deimos mission would have requirements and costs similar to that of the near-earth asteroid mission.

Table 16-1 gives an overview of the relative accessibilities of the lunar surface, selected earth-approaching asteroids, Phobos, Deimos, and the Martian surface. Measured in units of velocity interval (ΔV), the results may seem surprising at first glance: that selected asteroids, Phobos, and Deimos, are more accessible to the earth and less expensive in fuel to reach than the lunar surface. One result is more economic operations and earlier investment return, as expressed in Table 16-2. Both tables provide a framework for the description of missions discussed in the following sections.

Return to the Moon

The most actively pursued post-space-station goal in the United States is a return to the moon and the establishment of a lunar base for scientific exploration and resource processing. The two initial study teams, Science Applications International Corporation (SAIC, formerly SAI)[10] and the Arthur D. Little Company,[11] examined what a near-term lunar base would look like, what kind of science unique to the lunar environment would be carried out, how this would dovetail or conflict with the space station, how much it would cost, and what economic benefits may accrue.

Probably the most important lunar attribute is its soil. The raw lunar dust will be immediately useful in covering over the human habitats and laboratories to protect the astronauts from cosmic rays, solar flares, and potential military threats. A two-meter thickness of lunar dust provides the same protection against cosmic rays we enjoy here on earth. Simple heating of the lunar materials by a solar furnace can give us sintered blocks, tiles, pipes, glass, ceramics, and cement.

Table 16-2 deals with the returned mass of processed metals from lunar and other surfaces.

The long-term value of lunar materials comes from chemically processing their pure elements for end use on the moon or in space.[12] It is fortuitous that the raw lunar dust contains mostly metal and silicon oxides—when broken down, 40 percent oxygen, 20 percent silicon, and 20–30 percent metals. Astronauts could use lunar oxygen for breathing, food, and water, and to fuel rockets that go to and from the moon and between low-earth and geosynchronous (24-hour) earth orbits. Silicon would be fabricated into solar collectors and glass.

TABLE 16–1.
MISSION OPPORTUNITIES TO THE MOST ACCESSIBLE KNOWN EXTRATERRESTRIAL RESOURCES

Target	Launch Date (or Frequency)	Round-Trip Travel Time	Velocity Intervals ΔV (KM/SEC.)			
			Escape	Rendezvous (Land)	Depart	Total*
Lunar Surface	Frequent	≥7 Days	3.2	2.7	2.4	8.3
Martian Surface (Atmospheric Braking)	Every 2 Years	≥2 Years	3.6	≥1.0	≥5.6	≥10.2
Phobos and Deimos	Every 2 Years	≥2 Years	3.6	1.9	1.8	7.3
Phobos and Deimos (Mars Aerobrake)	Every 2 Years	≥2 Years	3.6	≤0.5	≤1.9	≤6.0
Asteroid 1982 DB	Sept. 2001	4 Months	3.2	5.8	4.3	13.3
Asteroid 1982 DB	Jan. 2001	2.0 Years	4.3	0.8	0.5	5.6
Asteroid 1982 DB	Dec. 2001	2.1 Years	4.5	0.5	0.7	5.7

*These figures assume aerobraking at earth.

TABLE 16-2.
ALLOWABLE RETURNED MASS OF PROCESSED METALS AS A FUNCTION OF BALLISTIC RETURN VELOCITY INCREMENT (ΔV)*

Return ΔV (km/sec.)	Tons of Returned Mass Per Ton of Invested Mass	Examples	Launch Date (or frequency)	Round-Trip Travel Time
0.1 0.2	100 49	1982 DB at special times	2000–01 (every two decades)	2 Years
0.5 1.0	19.2 8.9	1982 DB, 1943 Anteros, 433 Eros, and 1982 XB	Occasionally (for a given object, every decade or more)	2–3 Years
1.9	4.0	Phobos and Deimos	Every 2 Years	2 Years
2.4	3.0	Lunar Surface	Frequent	~7 Days
~5	1.0	Martian Surface	Every 2 Years	~2 Years
		1982 DB	Sept. 2001	4 Months
		Several Near-Earth Asteroids	Frequent	~1–5 Years

*Assume aΔV of about 5 km/sec for the trip out from low earth orbit. The same cryogenic rocket with exhaust velocity 4 km/sec is used for the return trip as for the trip out, the rocket is refueled with liquid oxygen (and possible liquid hydrogen) obtained at the asteroid, and the materials are aerobraked at the earth prior to landing or orbit insertion.

Metals would go into building large space structures, space stations, deep exploration vehicles, lunar base facilities, and radio telescopes. The key is becoming free of the earth's gravity well and using lunar materials as much as possible.

Chemists have devised a number of feasible schemes for extracting pure elements in lunar processing plants. Each process involves physical separations using solar or nuclear energy to heat or move around the materials; and chemical separations of the oxygen, silicon, and metals from their oxides through the use of reagents such as hydrofluoric acid. Depending on what products are needed, the processes involve a large number of complicated steps, but they are well-known from terrestrial experience.

The chemical processing plants would be self-contained units that recycle the reagents and churn out products whose mass is several times that invested in the processing machinery sent from the earth. In the SAIC scenario, a small pilot chemical plant weighing one ton could process ten tons or more per year of lunar materials into oxygen, metals, and glass.[13] The ten-to-one leverage of yearly production of useful products over the amount of mass invested in sending the processing machinery from the earth becomes important in the economics of becoming independent of the earth.

Not everybody agrees that the moon should be our next step.[14] The apparent lack of water on the moon would make it necessary to launch hydrogen from the earth or elsewhere for refueling and for life support. Moreover, the chemical separation of oxygen from metal and silicon oxide is more costly than extracting the oxygen by the electrolysis of water.

Some of these objections would be removed if ice is trapped in the permanently shadowed areas near the lunar poles. A lunar polar geochemical orbiter prospecting mission planned for the 1990s will probably answer the question. If the results are positive, we can envision a self-sufficient lunar settlement near the poles with a capacity for refueling all rockets that operate between low earth orbits and beyond. Full-time sunshine would become available to the lunar polar base from solar collectors on mountain tops.

Looking beyond self-sufficiency on the moon is the possibility of supplying earth orbital facilities with lunar manufactured materials. It costs twenty-two times less in energy to supply high orbits about the earth with lunar materials than with earth materials.[15] These will be the sites of large growing space industries, solar power satellites for providing energy to the earth, space settlements, large exploration vehicles, and fuel depots. One aspect of rapid growth to lunar self-sufficiency is the use of a long and skinny launching machine called a mass-driver. This electromagnetic motor will be able to accelerate lunar materials in recyclable buckets at accelerations greater than 1000 g, traveling from zero to lunar escape velocity of 2.4 kilometers per second. As a proof-of-concept, a tabletop mass-driver in a laboratory at Princeton University is now performing at accelerations close to those projected in the lunar scenario.[16] Also, the lack of any air resistance on the moon permits simple operations involving the high velocities and precise guidance required. Superconducting coils dipped in liquid helium carry electricity from the solar panels along the mass-driver track. The astronauts would set up the coils in key positions along the track to generate strong magnetic forces that both levitate and accelerate each bucket as it passes by, one every two seconds. After electronically fine-tuning the guidance, each bucket would slow down slightly and release a bag containing about two hundred grams of lunar material to escape the moon's gravity. The bags would form a continuous pipeline of material launched into space with eighteen hundred tons accumulated in free space after one year.[17]

When the lunar materials reach the lunar escape velocity and their guidance is fine-tuned by electromagnetic feedback systems, they drift out to a point 63,000 kilometers behind the moon, later to be caught by a cone-shaped receptacle about 100 meters across. This location named L_2, is one of the Lagrangian stability points in the earth-moon system, selected because the mass-catcher can stay in one place with the minimum amount of fuel. The material would then be pushed gently to a stable high orbit about the earth, where it would be processed into industrial products.

Meanwhile, the lunar facilities would grow as self-replicating machines fabricate more mass-driver components. One study showed it possible to start with the investment necessary to put 100 tons of machinery on the moon and after twenty years of machine reproduction to have an energy plant and manufacturing capability equivalent to the ability to manufacture 20 billion pounds (10 million tons) of aluminum per year.[18] The conclusion was that the technology is presently available to build such self-replicating machines and that the necessary development could be done in a decade or so.

The robots would in no way resemble R2D2 or C3PO immortalized in *Star Wars*. Instead, they would be "dumb" special-purpose machines that do simple redundant tasks as in an automobile assembly plant. After two years of replicating themselves and the components for the mass-driver, the mass-driver would grow to handle 100,000 tons of lunar materials per year almost entirely independent of what needs to come from earth. Within a few years, solar power satellites fabricated from lunar materials could deliver cost-competitive, environmentally-compatible electricity to the earth via a low-density microwave link, and large space habitats could be built at a small fraction of the cost required to launch these materials from the deep gravity well of the earth.

Essential to creating a lunar base is developing upper-stage rockets that could send personnel and cargoes beyond the low earth orbits that space shuttles and space stations will be confined to during the 1990s. These orbital transfer vehicles (OTVs) will tug payloads to and from 24-hour geosynchronous orbit and on trajectories toward the moon, asteroids, and planets. By 1996, NASA hopes to develop a fleet of reusable OTVs that would be periodically refueled at a space station gas station. The OTV is clearly a prerequisite to sending people anywhere beyond the space station.

The SAIC study[19] also considered developing lunar descent stages for landing cargoes and crew modules that can land on and take off from the moon. A crew of six to eight would be housed in a modular base that closely resembles the space station in size and layout. Included would be a laboratory module for examining lunar samples, monitoring health, and producing food; a habitat module; an unpressurized resource or utility module; a small pilot chemical plant for processing lunar materials; a connecting module with an observation dome and airlock with access to the lunar surface; two vehicles that at first serve as mass movers and later become lunar rovers. As in the space station, each module would be sized to fit in the space shuttle cargo bay and have dimensions of about fifteen feet in diameter and forty feet long.

A typical crew might consist of a commander/pilot, a mechanic/pilot, a technician/mechanic, a doctor/scientist, a geologist, a chemist, and a biologist/doctor. Crew members would rotate every two months with transfers of three or four of the crew occurring between OTVs and lunar excursion modules that rendezvous in low lunar orbit each month. This rather conservative scenario borrows technology from Apollo and the space station wherever possible and paves the way to rapid industrial growth in space.

The studies also describe experiments that would take advantage of the unique lunar environment. Using the moon as well as the earth as a stable platform for large antenna arrays, radio astronomy will open new vistas. The earth-moon distance of 384,000 kilometers will provide a baseline for resolving detail in radio sources thirty times greater than has been possible using the earth alone.

The search for extraterrestrial intelligence employing radio telescopes on the lunar backside will become vastly more productive because the radio chatter from the earth is blocked out. High energy astrophysics at large lunar sites will be able to take advantage of both a vacuum environment and the availability of abundant absorbing materials—a combination of features absent on the earth and in free space. Lunar geology, which barely began in the U.S. Apollo and Soviet Luna programs, could thrive at a permanent lunar base.

Scientists at the NASA Johnson Space Center are working on a twenty-five-year plan that culminates with a permanent lunar base. They have developed three scenarios: a science base, a production base, and a lunar self-sufficiency research base. The SAIC, Arthur D. Little, and NASA studies all show that lunar science and processing could begin for an investment of between $20 and $55 billion, which in current year dollars is less expensive than the Apollo program. This initiative would cost less than 1 percent of the U.S. federal budget spread over eight years of preparation and is less than NASA's current total budget.

It is hard to believe that more time has elapsed since anybody has walked the moon than the age of the entire space program at the time the first Apollo astronauts landed on the moon. One of those men, Buzz Aldrin, recently reminisced about that historical moment.[20] "It was just fifteen years ago," he said, "that I stood outside a cramped cockpit, looking out a triangular-shaped window and watched as my colleague, Neil Armstrong, made the first footprint on the surface of the moon. I heard him say, a second before anyone back on earth did, that this was 'One small step for man, one giant leap for mankind.'"

". . . Today, I believe it is time we make the giant leap a reality for mankind. . . . It's time to return to the moon—for good."

Near-Earth Asteroid Missions

Most studies on extraterrestrial materials processing have addressed the lunar soil. The main advantage is that the times of transit to and from the moon are shorter by two orders of magnitude than the lowest-energy travel times to Mars and to the earth-approaching asteroids. But if we envision the stream of material as a steady-state pipeline once the space mining and transporting systems are in place, the transit times and distance will not become a major cost factor. Long travel times and incomplete knowledge are presently making the asteroid option less attractive than the lunar case.

Several features of asteroid mining make it more advantageous than moon mining in the long term: their greater accessibility in terms of energy (measured by a lower required velocity increment ΔV for propulsion); higher abundances of precious metals; the availability of easily extractable free metals on some of the asteroids versus what appear to be only metal oxides on the moon; the probable availability of water, carbon, and other volatiles for refueling and life support; and full-time sunshine at the asteroid as an energy source.[21]

Approximately 20,000 asteroids with diameter greater than 200 meter (15 million tons) pass close to the earth over time. Once such asteroid of ordinary chondritic (i.e., the common meteorites) composition could supply more than 500 tons of platinum-group metals worth over $5 billion at today's market prices.[22] About 10 percent of these asteroids present opportunities for lower-energy (lower ΔV) round trips than travel to the lunar surface.

More significant are unique opportunities to send large tonnages of asteroidal material on earthward trajectories at very low energies. In at least one case,[23] the asteroid 1982 DB, the return ballistic velocity increment (ΔV) can be as low as 0.1 km/sec (Table 16-2), more than thirty times lower than escape from the lunar surface. When we consider aerobraking in the earth's atmosphere, the fraction of mass expended as fuel in a conventional rocket becomes only a small fraction of the total returned mass (Table 16-2). It follows that much larger masses can be returned from the asteroid than sent to the asteroid to do the processing.

Until recently, mission studies to earth-approaching asteroids have fallen into two categories: scientific reconnaissance of specific targets, possibly with the return of kilograms of sample;[24] and retrieval of $\geq 10^6$ tons of asteroidal material for end use in space.[25] A sample return mission requires as little as one space shuttle launch with a recoverable upper stage. The mass of material returned in the two cases spans eight to nine orders of magnitude.

The scale of initial extraction and retrieval of platinum-group metals from an asteroid is a geometric mean between these extremes: ≥ 100 tons. This scale is probably most realistic for the beginning of a commercial program in terms of available technology, operational cost, and initial benefit of $\geq$ \$1 billion.[26].

Mining and processing the most valuable materials appears to be the most cost-effective route to opening up the resources of space. Platinum-group metals refined at the asteroid and returned to earth qualify as the most likely first products. The \$5000/kg to \$10,000/kg value of these materials is comparable to the cost of building and transporting space hardware to geosynchronous orbit and beyond. This suggests that if the weight of returned metals is comparable to that of the hardware required to process and retrieve them, a commercial program could be justified.

Alternatively, if production significantly exceeds the mass of the processing machinery sent to the asteroid, then it would be cost-effective to deliver low grade ores. We have seen that special opportunities exist to send large tonnages to the earth at ΔV values ~ 100 times less than that required to send equipment from the earth's deep gravity well to the asteroid. This fact of physics suggests maximizing the productivity of useful materials per unit mass of the mining equipment. If the same rocket is used for sending the mining payload out and bringing the metal in, the ratio of deliverable metal mass to invested mining mass should be as large a number as possible—on the order of 100 in at least one case. Using the rocket equation for a cryogenic rocket, Table 16-2 shows the processed material masses that could be returned as a function of the return velocity increment and the mass invested in the processing machinery.

In all cases, we require a system that could process very large throughputs of raw materials from the asteroid. For example, producing 100 tons of platinum-group metals at 10 parts per million demands that 0.1 tons per second of feedstock be processed over three years. While there is plenty of material available to do this, a conventional terrestrial "sand-and-gravel" approach requires that several thousand tons of machinery and powerplant be sent up at the asteroid to crush, grind, separate, and further refine the material.[27] Only by bootstrapping asteroidal materials and equipment self-replication could this traditional extraction method be viable, an option that is probably farther in the future.

On the other hand, the zone refining of varying purities of the needed asteroidal metals[28] permits the use of lightweight solar concentrating mirrors as the source of energy that would heat a portion of the asteroid above the melting point of the iron-nickel mixture that predominates in the composition of many candidate asteroids. A temperature of $\sim 1600°C$ is adequate to separate the metal fraction from any silicate matrix. The next step is to do the zone refining: successively drop and raise the temperature over a range of $\sim 100°C$ to create liquid-solid phase separations and allow for repeated increases in the concentration of the trace precious and strategic metals with respect to the more dominant metals. As the material begins to heat, volatiles, if present, could be collected on the shady side of the asteroid for rocket fuel for the return trip.

Independently, M. J. Gaffey and J. Green (unpublished) have investigated the equipment requirements for zone-refining for precious and strategic metals on an asteroid. In Gaffey's scenario, a solar furnace cavity and associated equipment weighing ~ 100 tons (with another ~ 100 tons for moving material, etc.) process 2.4 tons per second of raw material through the first of 15 to 30 cycles of zone refining. The concentration of trace metals in the final sample (principally gallium, germanium, and platinum group) increases to about 80 percent after 30 cycles. About 4,000 tons per year of material are produced by this method, with a market value between \$1.5 and \$5 billion per year. The ratio of produced material to invested mass in this scenario (about 20 per year or 100 after 5 years) is the same order of magnitude as the ratio derived in Table 16-2 for the ballistic (low energy, coasting) return of materials from asteroid 1982 DB in 2001. In addition, crews and small payloads can be sent to 1982 DB with as low a round trip time as four months in 2001.[29]

In 1977, a group of scientists and engineers at the NASA Ames Research Center developed scenario in which a mass driver, assembled in space, retrieves a 0.5 million to 2 million ton piece of asteroidal material through a velocity increment (ΔV) of 3 km/sec in about five years. These velocity increments are comparable to those to and from the lunar surface, but no "soft landings" would be required and solar energy would be continuous at the asteroid. Many such candidate asteroids are within reach of earth-based telescopes in ongoing search programs.

The NASA study included trajectories using gravity-assist by planetary encounters.[30] Two other papers describe in detail the velocity increment requirements for favorable round-trip missions to currently known candidates and probable future cases,[31] and an assessment of asteroidal resources and recommendations for expanding the search program, follow-up for orbital determination and chemical classification, and identification of precursor missions.[32]

The scenarios included the best (then) known case (Bacchus) with gravity assists from the earth, Venus, and moon, and a likely hypothetical case, given an increase in the asteroid discovery rate and improved mission analysis techniques. A parametric study of the scenario identified the most significant variables in comparing the economics of transport into a stable high orbit, for asteroidal, lunar, and terrestrial materials.

The study concluded that asteroidal resources are often less expensive than lunar materials and are far less expensive than terrestrial materials to place in high orbit. The group recommended a research and development program designed to provide technological readiness for asteroid retrieval. Such a program would include an expanded search program and a new start on precursor missions to prime candidates selected for their accessibility and class of chemical composition. These recommendations were corroborated in a second NASA-funded workshop in La Jolla, California.[33]

Human Missions to Mars and Its Moons

Probably more than any object in the Solar System, the planet Mars beckons humanity. We are living in a moment of history in which the Red Planet is a focus for a number of seemingly unrelated events and goals: exploration, improved East-West relations, industrial growth, spiritual awareness, and perhaps the first signs of extraterrestrial life.

For nearly a century the question of life on Mars has opened and closed many times: Percival Lowell's canals, William Sinton's organic molecular spectra, and the U.S. Viking Lander's ambiguous findings in 1976 by its biology experiments have all seen their day. The question has been reopened by a second scientific look at some tantalizing features photographed by the Viking orbiter that resemble gigantic eroded pyramids and a mile-wide human face.[34]

Indications from all directions point to a Mars renaissance. With the space station now authorized in the United States, and with Soviet initiatives toward Mars, formerly reticent NASA officials are now expressing as very real possibilities human trips to the moon and Mars early in the twenty-first century. Politicians in both the United States and Soviet Union are echoing these sentiments, with an emphasis on Mars.

How this magnificent feat will be first done is open to debate. The lowest cost approach would be to go first to the two moons of Mars, Phobos and Deimos[35]. The potato-shaped asteroid-like bodies measure about 25 and 14 kilometers across and were photographed extensively in the Mariner and Viking missions to Mars.

Because of their low gravities, no fuel-consuming soft landings or blast-offs are needed as in the cases of Mars and the moon.[36] In terms of energy and rocket fuel they are closer than the lunar surface! And they are likely to contain large amounts of easily extractable water bound in the minerals that could be stockpiled for refueling and life support.[37] The door would open for rapidly growing space industrialization.

If Phobos and Deimos were moons of the Earth we would be there by now. Circling Mars in relatively low orbits, each moon would provide a stable platform from which astronauts could teleoperate unmanned Mars rovers and survey landing sites within a light second, not light-minutes of feedback time needed from Earth.[38]

S. Fred Singer[39] presented a scientific rationale and plan for the manned exploration of Phobos and Deimos (PhD) and subsequent unmanned exploration of the Martian surface. The PhD mission, wrote Singer, would provide a number of advantages over either sending unmanned rovers or people directly to Mars first: tele-metered control of rovers from the Mars-facing sides of the satellites will shorten the response time from 50 minutes to less than 0.2 seconds; surface and subsurface Martian samples can be recovered at disparate sites by remote control and examined in a PhD laboratory rather than having to resort to a sophisticated laboratory on Mars or to run the risk of back-contamination by direct sample return to the earth; PhD-controlled surface rovers allow for sequential exploration of Mars at several locations rather than direct shots that may miss the most interesting spots (Viking and Apollo suffered from this prob-lem); unmanned landings are easier, less costly, safer, and could be done much sooner; and we could directly sample the moons to deter-mine their history and fate, gaining valuable insight on the origin and evolution of small bodies in the solar system.

Adelman and Adelman[40] have presented advantages of using Phobos as a base for gravity wave astronomy, radio astronomy, astro-metry, and the study of asteroids and comets. Also, a Mars-pointing telescope on Phobos will permit resolutions an order of magnitude or more greater than that of the Viking orbiters (C. Sagan, private communication). The Soviets have launched an unmanned reconnais-sance mission to Phobos in 1988. Clearly, the incentive to explore and utilize the satellites of Mars is increasing.

O'Leary[41] has investigated the resource potential of the Martian moons and their accessibilities. The very low mean densities of both moons (~ 2 g/cm^3) and their low reflectivities are strong indicators of a chemical content similar to that of the volatile-rich carbonaceous chondrite meteorites.[42] Easily extractable water is likely to be found on both moons. Studies described in the previous section examined the rationale and mission scenarios for setting up resource recovery operations on selected earth-approaching asteroids during special launch opportunities. The Martian moons can also be considered as earth-approaching asteroids with mining potential, with the added features of more accessibility in two-year launch intervals, the prox-imity of Mars, and the likely availability of volatiles. The most ac-cessible known asteroid, 1982 DB, offers slightly better opportunities at decade intervals, but missions in the intervening years deteriorate rapidly.[43]

Earlier studies,[44] have also pointed out that, in terms of the energy (and cost) of setting up an oxygen-extracting plant on a non-terrestrial body, the presence of water and other volatiles is far prefer-able than the chemical processing of metal oxides that predominate in the lunar soil and many meteorite classes. Carbonaceous materials also make available hydrogen, carbon, and nitrogen—elements that will become useful for refueling and for life support.

In the long run, the absence of these materials on the moon or asteroid would necessitate expensive resupply from the deep gravity well of the earth. Cordell[45] has estimated that the energy cost of delivery of water to lunar base from Phobos and Deimos would be approximately three times less than that from the earth's surface.

In Table 16-1 are estimates of the ballistic velocity increments (ΔV) required to rendezvous with Phobos, depart, and return to the earth.[46] Table 16-1 shows the results for average launch windows from earth every two years, assuming Hohmann transfer (minimum energy) ellipses between circular, coplanar orbits. Primarily because of the eccentricity of Mars's orbit, the figures for each launch window will vary up and down from the values reported here, but they are representative values for mean opportunities, as indicated by cross-checking with analyses of specific mission opportunities.[47]

The total mission velocity increment for Phobos, Deimos, and other targets can be divided into three principal maneuvers: escape from low earth orbit onto the transfer ellipse, rendezvous with Phobos/Deimos, and return on a transfer ellipse to the earth. Fuel savings would result from "aerobraking" at both Mars and the earth. These maneuvers use the top of the atmospheres of both bodies to eliminate excess incoming hyperbolic velocities and to allow for capture within their gravitational fields.

Current studies of orbital transfer vehicle (OTV) systems to be used in the post-1996 time frame show an aerobrake capability.[48] Because the velocities of atmospheric encounters projected for the earth are greater than those at Mars, no significant design requirements need be added to a spacecraft with roundtrip capability to Mars.

One result that may be surprising at first glance is that even without aero assist at Mars, the moons of Mars are more accessible to the earth at biennial opportunities than is the moon of the earth. The chief difference is in the requirement to soft-land payloads on the lunar surface. With meager escape velocities of 11 and 6 meters per second for Phobos and Deimos, the impulse required to soft land on these Moons is negligible. Required maneuvers more resemble rendezvous and docking with a spacecraft rather than impulsive blasting into and out of a gravity well. As in the case of the asteroids, the PhD missions permit low impulse propulsion for the entire trip, opening the possibility of using solar-electric, mass-drivers, solar sails, and tethers (use of cables to transfer angular momentum) as the sources of propulsion.

The only advantages the moon seems to offer are its proximity and launch window frequency: days versus months to years. But once it has been established that humans can survive in space over long periods of time, and once mining and exploratory operations begin and a pipeline of extraterrestrial materials starts to flow, it will probably not make much difference how close the source of materials is to the earth. More significant will be the energy required to process and transport them, and Phobos and Deimos provide an advantage.

Two Apollo circumlunar flights were attempted before a lunar

landing. Likewise, politics, safety, and economics point to going to
Phobos and Deimos before attempting to land humans on the surface
of Mars. The recent Office of Technology Assessment report on the
U.S. space program[49] refers to missions "to the vicinity of Mars"
rather than Mars itself. Former Apollo astronaut and U.S. Senator
Harrison Schmitt[50] foresees a possible Soviet human mission to
Phobos by 1992. The 1988–89 Soviet Phobos probes reinforce this
interest.

Phobos and Deimos offer more than enough materials to indus-
trialize space: within their volumes they contain the equivalent of 10
million 100-meter diameter (1 million ton) objects! The milligravity
environment of the Martian moons will permit certain industrial
operations that may be more easily carried out than either in complete weightlessness or under the influence of a significant gravitational field. Dust, equipment, and people will not float away; yet
structures need not be built to withstand a planetary or lunar gravity.

Table 16-2 expresses the ratio of returned to invested mass from
these objects with the ground rules of each mission stated in the
footnote. It is clear from inspecting this table and from the preceding
paragraphs that Phobos and Deimos offer a unique blend of exploration, resource processing opportunities, accessibility, and frequency
of launches. These factors combine to create a powerful incentive to
conduct a detailed analysis of mission scenarios.

Singer[51] has proposed Deimos to be the first manned destination,
primarily because the outer moon's orbital period is nearly syn-
chronous with Mars's rotation, allowing for more continuous tele-
metered operation of surface rovers. Also, Deimos is in the line of
sight of locations at higher Martian latitudes.

Phobos, on the other hand, provides some advantages. In terms
of ΔV, it is more accessible to the Martian surface. Its position deeper
within the Martian gravity well grants it slightly more accessibility
for circularizing after an aerobraking encounter, but normally
slightly less favorable return-to-earth circumstances. And Phobos
provides a higher-resolution view of the Martian surface. By the time
a manned mission is launched, we are likely to have more knowledge
about the composition of Phobos than that of Deimos. A confirmation
of water in the forthcoming Soviet mission to Phobos might be the
deciding factor.

In reality, it will be desirable to visit both moons. From an
operational point of view, redundancy is desirable. In the long run the
two satellites could serve complementary roles—Deimos for astronomy, planetary coverage, reconnaissance, and control of Mars rovers;
and Phobos as a site for volatile mining if water is found there, sorties
to and from the Martian surface, and close-up surveillance of interesting sites. It may also be convenient for Soviet bloc nations to focus
on one moon and western countries on the other in a cooperative
effort. The ΔV between the two moons is a convenient 747 meters per
second for two-burn Hohmann (minimum energy) transfers.

The next logical step is to define mission scenarios, list scientific objectives, and perform cost and performance trades with other options, such as the direct manned mission from earth to the Martian surface. One concept includes a Venus swing-by with opportunities in 1999, 2001, and 2003, where the PhD staytime is two months and total mission time between six hundred and seven hundred days.[52] Features of the scenario would include lower overall cost and safer operations than a manned Mars landing, more quality Martian science by sequential exploration using unmanned rovers at disparate sites telemetered from PhD, collecting and launching Mars samples by unmanned vehicles and analyzing them in a PhD laboratory, sampling the Martian moons themselves, and creating the infrastructure for processing fuels from PhD volatiles or oxides, with rapid expansion thereafter. The only apparent missing feature of this first mission is the manned exploration of the Martian surface. If, for political reasons, a manned Mars landing is warranted, the PhD scenario could include a sortie to the Martian surface in a small vehicle at some additional cost.

Many of the same ground rules for analyzing other manned missions (lunar bases, Mars landing missions, and near-earth asteroid missions) apply to the PhD case. Two studies carried out principally by the Schaumberg, Illinois, group of Science Applications International Corporation (SAIC) provide the basic logistical requirements and costing assumptions for an initial lunar base,[53] lunar base reconnaissance mission,[54] a near-earth asteroid mission,[55] and a first manned Mars landing.[56, 57]

In the last case, the investigators assumed a dual launch in 2003, aerocapture at both Mars and earth, a four-man crew, 30-day stay time, and minimal new technology. Their overall cost estimate of $39 billion (1984 dollars) could be reduced by a factor of two or more by eliminating the need for developing a man-rated Mars lander, which includes the fuel needed for an ascent stage to get out of Mars's gravity well (more shuttle flights, more OTVs). The manned Mars landing mission would more resemble a one-shot Apollo flight than the scenario of a sustained operation at a small base on PhD that would be manned initially and visited periodically for expanding the materials processing capability and upgrading the exploration program.

The economics of establishing and growing a PhD base appears to be more immediately feasible than those required to start and sustain a base on the Martian surface deep within its gravity well. Table 16-2 shows that the ratio of returned to invested mass for delivery in free space is about four for Phobos and Deimos and only one from the surface of Mars. Phobos and Deimos will provide among the first of our nonterrestrial material resources.

The case for Phobos and Deimos enhances the case for Mars. We appear to have a situation of serendipity in science and economics. The PhD scenario requires new legitimacy previously not given it because of an inherent familiarity with doing things at the bottom of deep gravity wells. But the laws of physics and practical economics dictate we take a second look at our future priorities in space exploration.

I recommend that mission planners investigate the PhD options in as much detail as is now being given to manned Mars landings and lunar bases. These tasks should not be difficult because methods for determining most of the assumptions, mission parameters, and logistical requirements are common to all scenarios. In my opinion, the results will lead not to an either-or proposition about destinations involving irrevocable dead-ends, but a synergistic blend of missions that will grow rapidly. Phobos and Deimos appear to be at the focus of the initial step from which everything else will logically follow. The essence of Martian exploration and space industrialization using non-terrestrial resources is embodied in the fascinating moons of Mars.

Implications

Whichever scenario emerges first, second, third, or fourth for human missions (lunar base, PhD mission, Mars landing, asteroid mission), it is clear that the human expansion across the Solar System will accelerate. Space tourism, solar power satellites, ships of exploration, space colonies, bases on Mars, the other planets and satellites, terraforming, huge agricultural areas in space, and asteroid mines will become part of the emerging multi-trillion industry in space. The economic, scientific, and international political incentives are too compelling to be ignored.[58] It is likely we will see the widespread presence of people in space, on other planets, their satellites, and the asteroids by the mid-twenty-first century. The kinds of individuals who will take the first one-way trips into space will probably more resemble Pilgrims than astronauts with "The Right Stuff."[59]

The implications will be profound. Space settlements will provide new ecological niches that will breed individual human diversity and accelerate their evolution in various directions we cannot now anticipate. We may soon encounter extraterrestrial intelligence. We may be living in a time no less significant than when our fish ancestors first crawled onto land.[60] But the implications also apply to the whole as well as to individuals. As the biosphere of the Earth evolves as its own superorganism (according to the Gaia hypothesis[61]), so will the human settlements themselves evolve as organisms in space. Evolution will proceed simultaneously on many different levels.

The opportunity for international cooperation is unprecedented. By a margin of more than ten to one, leading American scientists, theologians, environmentalists, and Congressional staffers in a recent opinion poll preferred to cooperate with rather than upstage the Soviets in a human voyage to Mars.[62]

In the United States, President Reagan has signed into law the Matsunaga bill, unanimously passed by Congress, that provides for U.S.-Soviet cooperation in space. Senator Matsunaga sees an international human mission to Mars as the "most stirring undertaking in human history" that would "herald the birth of a new transcendent age."[63]

"Suppose the people of the Earth," wrote astronomer Carl Sagan,[64] "are one day fortunate enough to discover new leaders in Washington and Moscow dedicated to a new beginning; and to seal that new beginning they embark on a dramatic joint enterprise— something like an Apollo program but with cooperation, not competition . . . could we muster a mission to Mars with human crews for the sorts of money repeatedly allocated for weapons systems on earth? Astonishingly, the answer seems to be yes."

"The earth is a cradle", wrote the Russian schoolteacher and visionary Konstantin Tsiolkovsky at the beginning of this century, "but man cannot stay in the cradle forever." With the coming of the new millennium and 100 years after Tsiolkovsky's prophetic statement, we are taking our first tentative steps away from our fragile, finite planet. It is exciting to be living in these times. Space is a metaphor for human transformation into the New Age.

NOTES

1. Lunar Base Working Group Report to NASA. Meeting at Los Almos, April 1984.

2. "Lunar Bases and Other Space Activities in the 21st Century." NASA/NAS Symposium, Washington, D.C., Oct. 29–31, 1984, in press.

3. M. Washburn, "The Moon—A Second Time Around?" *Sky & Telescope*, March 1985, 209.

4. Science Applications International Corporation, "A Manned Lunar Science Base: An Alternative to Space Station Science?" Report No. SAI-84/1302 to the National Science Foundation, Jan. 19, 1984.

5. H. Simmons, "Four More Years: Aerospace Wins Big," *Aerospace America*, Jan. 1985, 14.

6. S. Fred Singer, "The PhD Proposal: A Manned Mission to Phobos and Deimos"; *The Case for Mars*, Science and Technology Series, American Astronautical Soc. Volume 57, Univelt, Inc., 1984, 39–65; also "The PhD Project in Perspective," The Case for Mars II Conference, Boulder, Colorado, July 1984, in press.

7. B. O'Leary, "Phobos and Deimos (PhD): Concept for an Early Human Mission for Resources and Science," *Space Manufacturing 5*, New York: American Institute of Astronautics, (1985); *ibid 6* (1987); *Mars 1999: Exclusive Preview of the U.S.-Soviet Manned Mission*, Harrisburg, Pa: Stackpole Books (1987); "Mars 1999: Concept for Low-Cost Near-Term Human Exploration and Propellant Processing on Phobos and Deimos," The Case for Mars III, American Astronautical Society (1989, in press); "International Manned Missions to Mars and the Resources of Phobos and Deimos," presented to the International Astronautical Federation, Bangalore, India (October 1988).

8. B. O'Leary, "Mining the Apollo and Amor Asteroids." *Science* 197 (1977): 363– 366; also in *Space Manufacturing Facilities*, New York: American Institute of Aeronautics and Astronautics, Vol. 2 (1977): 157.

9. Science Applications International Corporation, "Manned Lunar, Asteroid, and Mars Missions." Report to the Planetary Society, Pasadena, California, Sept. 1984.

10. SAIC, *Op cit.*, reference 4.

11. H. Simmons, *Op cit.*, reference 5.

12. G. O'Neill, *The High Frontier*, New York: William Morrow and Company, 1977; B. O'Leary and G. O'Neill, "Space Manufacturing, Satellite Power and Human Exploration." *Interdisciplinary Science Reviews* 4, no. 3 (1979): 193–203.

13. *Ibid.*

14. B. O'Leary, "Lunar Base: Many Experts Say It Is Our Next Stop." Science Digest, Jan. 1985, 59–63.

15. G. O'Neill, "The Colonization of Space." *Physics Today*, Sept. 1974, 32–40.

16. L. Snively and G. O'Neill, "Mass Driver III: Construction, Testing and Comparison to Computer Simulation," Princeton, New Jersey: Space Studies Institute Mass Driver Research Project, 1984.

17. G. O'Neill, G. Driggers, and B. O'Leary, "New Routes to Manufacturing in Space." *Astronautics & Aeronautics*, Oct. 1980, 46–51.

18. *Ibid.*

19. SAIC, *Op. cit.*, reference 4.

20. B. Aldrin, "Let's Return to the Moon for Good," *Los Angeles Times*, July 22, 1984, Part IV, 5.

21. B. O'Leary, *Op. Cit.*, reference 8.

22. *Ibid.*

23. N. Hulkower, "Opportunities for Rendezvous and Round Trip Missions to 1982 DB based on Update Elements," Jet Propulsion Laboratory Interoffice Memorandum 312/82.3—2046, June 1982; R. Staehle, "Finding 'Paydirt' on the Moon and Asteroids." *Astronautics and Aeronautics*, Nov. 1983, 47.

24. A number of reports submitted to NASA's mission plan between 1977 and 1984 describe details for these missions by the JPL Advanced Projects Group, N. Hulkower, R. Staehle, and J. Wright and by Science Applications International Corporation, J. Niehoff and A. Friedlander.

25. B. O'Leary, *Op. cit.*, reference 8.

26. *Ibid.*

27. *Ibid.*

28. For a detailed discussion and bibliography on zone refining techniques, see G. Atwood, "Developments in Melt Crystallization" in *Recent Developments in Separation Science*, ed. N. N. Li, Boca Raton, Florida: CRC Press, 1972.

29. A. Friedlander, Memorandum to NASA, Code EL-4, Advanced Studies Manager. Science Applications, Inc., Schaumberg, Illinois, Nov. 1982.

30. *Ibid.*

31. D. Bender *et al.*, "Round-Trip Missions to Low Delta-V Asteroids and Implications for Material Retrieval," *Space Resources and Space Settlements*, NASA SP-428 (1979): 161–172.

32. M. Gaffey, *et al.* "An Assessment of Near-Earth Asteroids," 191–204.

33. J. Arnold and M. Duke, Summer Workshop on Near-Earth Resources. NASA Conference Publication 2031, 1978.

34. V. DiPietro and G. Molenaar, *Unusual Martian Features*. Third Edition, 1982; R. Hoagland, "Preliminary Report of the Independent Mars Investigation Team: New Evidence of Prior Habitation?" Berkeley, California, 1984; R. Hoagland, *Monuments on Mars*, Berkeley: Free Press, 1987; M. Carlotto, *Applied Optics* (1988, in press).

35. B. O'Leary, *Op. cit.*, reference 7.

36. *Ibid.*

37. *Ibid.*

38. S. F. Singer and B. O'Leary, *Op. cit.*, references 6 and 7.

39. *Ibid.*

40. S. Adelman and B. Adelman, "The Case for Phobos." Boulder, Colorado: *The Case for Mars II*, July 1984, in press.

41. B. O'Leary, *Op. cit.*, reference 7.

42. J. Veverka, *et al.*, "The Puzzling Moons of Mars," *Sky & Telescope*, Sept. 1978, 86.

43. B. O'Leary, *Op. cit.*, reference 7.

43. *Ibid.*

44. *Ibid.*

45. B. Cordell, "The Moons of Mars: Source of Water for Lunar Bases and LEO." Washington, D.C.: *Lunar Bases and Space Activities in the 21st Century*, NASA/NAS Symposium, Oct. 29–31, 1984, in press.

46. B. O'Leary, *Op. cit.*, reference 7.

47. S. Hoffman, and J. Soldner, "Concepts for the Early Realization of Manned Missions to Mars," Boulder, Colorado: *The Case for Mars II*, July 1984, in press.

48. G. Walberg, "Aeroassisted Orbit Transfer Window Opens on Missions." *Astronautics & Aeronautics*, Nov. 1983, 36.

49. Office of Technology Assessment, "Civilian Space Stations and the United States Future in Space," OTA report to Congress, 1984, 13.

50. H. Schmitt, "A Millennium Project: Mars 2000." Washington, D.C.: *Lunar Bases and Space Activities in the 21st Century*, NASA/NAS Symposium, Oct. 29–31, 1984, in press; also cited in *Space Business News*, November 5, 1984, 6.

51. S. F. Singer, *Op. cit.*, reference 6.

52. SAIC, *Op. cit.*, reference 4.

53. SAIC, *Op. cit.*, reference 9.

54. *Ibid.*

55. *Ibid.*

56. *Ibid.*

57. S. Hoffman and J. Soldner, *Op. cit.*, reference 47.

58. B. O'Leary, *Op. cit.*, reference 7.

59. B. O'Leary and J. Gabrynowicz, "From Plymouth to Phobos: The Self-Organizing American Experiment." The LIberty Fund Conference on Self-Organization, November 2–3, 1984, Miami, Florida, in press; also the Society for General Systems Research, Annual Meeting, Los Angeles, May 1985, to be published.

60. B. O'Leary, *The New Reality: A Scientist's Perspective on Individual and Planetary Transformation*, a book in preparation.

61. For a detailed discussion of the Gaia Hypothesis, see P. Russell, *The Global Brain.* Los Angeles: J. P. Tarcher, 1985, 31.

62. Preliminary results of an opinion survey carried out by Professor J. Miller, Director, Public Opinion Laboratory, Northern Illinois University in 1984.

63. Senator S. Matsunaga, "First Word," *Omni Magazine*, Dec. 1984, 6.

64. C. Sagan, "The Case for Mars," *Discover Magazine*, Sept. 1984, 26.

17

Comment on Brian O'Leary

JAMES E. OBERG

"As for me, my task is not to foresee the future, but to enable it"
—Antoine de Saint-Exupéry

Brian O'Leary's vision and imagination have never been known to be restricted by the gravity that holds our bodies fast to the surface of our birth world. His mind roams the Solar System, and in its roaming opens trails for others to follow and, each in his own right, extend.

His blueprint for human expansion into the Solar System is a sound one, and the edifice of a future spacefaring civilization constructed along the lines O'Leary delineates would be a thrilling, inspiring structure. Yet few space prophets now at the end of the second millennum could be shortsighted enough to imagine the mid-third-millennium "humanned" Solar System will actually turn out to look as we can imagine it. No, all we space prophets can hope for is to produce "existence proofs" of the feasibility and desirability of such extraterrestrial civilizations. The actual constructions must be left to others—inspired it is to be hoped by our visions, but armed with their own special plans and skills that are today still unimaginable.

Hence it is of no consequence that I might quibble with some of O'Leary's suggested techniques, since the technology base is constantly burgeoning, and since O'Leary himself has shown the classical space prophet's traditional open-mindeness about new "tricks of the spaceflight trade." Based on past experience, we are confidently able to *expect* such breakthroughs while not being able to specifically *predict* what they will be. The entire history of human exploration and discovery on this planet teaches that lesson, and this next phase cannot be expected to be different.

192

For example, I predict that a surprising "sleeper" spaceflight technology of the next half-century will be "tethers," long/strong cables connecting space vehicles in orbit. The "rope tricks" that can be accomplished involve momentum exchanges that at first look like magic, but which must—and will—become the "common sense" of future space operations. Such tethers will be complemented by the "mass drivers," that O'Leary mentions, electromagnetic catapults that a decade ago were the stuff of visions (and the jury-rigged hardware of specialists assocaited with O'Neill's "Space Studies Institute"), but that today are the centers of bountifully funded anti-missile research projects (where they are called "rail guns"). Such research could enable the engineering advances that transform lunar mining and asteroid shoving from the realm of space visionaries to the even more exotic universe of engineering development and funding.

Sadly, O'Leary still speaks in earthbound metaphors when he describes three *separate* goals for future activities, that is, the moon, the asteroids, and Phobos-Deimos. To treat them thusly is to bait the trap of "either-or," or the sequential, linear style of thinking characteristic of planetbound cultures. It is the commonality of these goals, not their differences, that needs to be stressed. Certainly the hardware is at least 80 percent common, and possibly more so if developed with sufficient long-range thinking and foresight. Buy one, we should be telling today's carthers, and you can get the second for 20 percent of full price, and the third for 10 percent. "Cheaper by the dozen" should be the motto of would-be asteroid miners. Such statements are both true and useful, and might make a good party line for space enthusiasts of the crucial next decade.

If one particular rock (or set of rocks) needs to be picked out, then I opt for Phobos-Deimos (and leave Mars for later, perhaps much later). This PhD strategy, championed by Fred Singer, could be the key (or the fuel depot) to the Solar System, as I came to realize during careful examination of the logic, over the past few years.

The "Mars Underground" made famous by two seminal private colloquia in Boulder, Colorado, has now surfaced and has been preempted by the space industry and bureaucracy.

Very well; it is the fate of the farsighted to be denounced as crackpots right up until their ideas become so widespread that their pioneering primacy is forgotten. Let it happen again: the "Phobos Underground" has already been conceived, and a newsletter of the Houston Chapter will soon appear. Its motto is:

$$H_2O@\$1/gm$$

That is, water at a dollar per gram. If Phobos (and/or Deimos) can support such refining processes, they open the literal high road to the asteroids, trans-Jovia, and even some interesting fossil comets.

The intellectual and conceptual leap of mining the asteroids is a great one, with the mental gulfs nearly as wide (in terms of difficulty) as the physical ones. To make the concept more "down home" and less "gee whiz," some authentic (and very useful!) items of human history can be applied.

For example, human beings have been mining the asteroids since the dawn of civilization. The first metals beaten into tools and weapons were meteorites, chips off of asteroids that accidentally fell to earth and were retrieved by ancient humans. Even today, some of the richest metal ores in the world are being mined at Sudbury, Ontario, only recently recognized as a two-billion-year old "astrobleme" ("star wound," or fossil crater). The dispute continues over whether the metal there is actually a physical trace of the miles-wide asteroid or merely native ores that belched up from deep inside the planet's mantle when the overlying crust was blasted aside. One advocate of the exogenous theory issued the quip that such a hypothesis called for not only pennies from heaven, but nickel too.

The moral of the story is this: future prospecting among the asteroids would be only a more efficient way of conducting a traditional human activity, by exploiting resources first where they fall and later by pursuing them to their points of origin, whether upstream, upwind, or up-gravity.

In any case, such spacefaring activities must at first be conducted by organized human collectives such as governments or transnational corporations. And here, in the issue of international cooperation (particularly U.S./USSR cooperation), I feel I must sound a depressingly cautionary warning bell. The price we may have to pay for getting the Soviets to join with us on the road to Mars may be far higher than the cash value—if any—we save in our space budgets.

An examination of the cast of characters in the current "together-to-Mars" chorus reveals some distressing motivations and proposed "bargains." It may look like a ticket to the planets, but it might prove to be a "veto" instead.

As I testified to the National Commission on Space during their hearings in Houston recently, the arguments for joint U.S./USSR manned missions are far more political than practical. Some of the most outspoken advocates of such cooperation for a manned exploration of Mars are the very same people who have been bitterly opposed to manned space flight in the past. They are now engaged in an extremely transparent political ploy, embraced (if not originally conceived) in Moscow: "Give up 'Star Wars' and we'll give you the universe!" The Soviets have been holding hostage the question of future cooperation, based on an American repudiation of advanced missile defense systems; some of the loudest and best-known advocates of "space cooperation" are academics and politicians who have been consistently pushing the anti-SDI doctrine for a long time. So their recent embrace of U.S./USSR cosmic togetherness rests on motivations that may not be reliable over the long term needed to initiate, fund, develop, build, and fly such missions. At any point along that years-long process, if their ulterior goals of U.S./USSR relations are satisfied through some other means, they are liable to abandon the joint manned space project with alacrity, and resume their former opposition to such projects. Their current support contains the seeds of a built-in downstream betrayal: by agreeing now to such an alliance, we would probably be giving eleventh-hour veto power over the project to people who are known to hate it, and to a nation known to hate us.

That anxiety may be grounds for some cynicism, and some caution, in enthusiastically rushing into such an alliance!

As a longtime advocate of expanded U.S./USSR space cooperation, I speak with some moral authority on this subject. I was publicly for a Shuttle/Salyut rendezvous when it was considered absurd, even insane. I have vigorously denounced the anti-cooperation myths such as "they stole our space secrets" (no sign of that) or "they're copying our shuttle" (a nitwit notion). If then I utter Cassandra-like warnings about proceeding with too high hopes, my motivations are founded on my sincere enthusiasm for fulfilling the space expansion blueprint O'Leary (and, before him, numerous other experts including myself) has laid out.

One hope I have held for "Shuttle-Salyut" is that it may help defuse the current torrent of venom issuing forth from Moscow on the American space program in general, on the Space Shuttle in particular, and on some individual astronauts by name. My deep-seated anxiety is that such a smear campaign may be pushing the Soviets to "talk themselves into a corner" from which their only feasible reaction to a polar orbit DoD shuttle mission may be warnings, threats, and ultimately force. The rhetoric is mounting to pre-justify a "Korean airliner in space" tragedy, I fear.

Agreement on a joint mission might just defuse such inflammatory talk.

But we cannot draw too close an analogy with the Apollo-Soyuz mission a decade ago. That project, after all, was a symptom, not a cause, of diplomatic detente. U.S./USSR relations had been improving, and the joint mission was commissioned to illustrate a motion that was already under way. The ASTP project did not *cause* such "good feelings" (the "beau geste" theory of diplomacy), but merely reflected them. And by the time the joint mission actually occurred, it was already an anachronism, with Soviet tanks rolling into Saigon, Soviet-sponsored coups in several African nations, planning well along for the "painless" takeover of Afghanistan, and so forth.

In another metaphor, the robin is the harbinger, not the cause, of spring, and Apollo-Soyuz was the "robin" for detente, even if it did not arrive until well into autumn! It has become enshrined as a myth of what can be accomplished if both nations are not hostile, but as with all political myths (and sausages, and laws), one should not observe too closely the processes by which they are created!

What can be done besides despair? Let me suggest we approach the problem as engineers, not politicians. The latter would prefer to build a single spacecraft with every first bolt produced in Detroit and every second bolt in Dnepropetrovsk. But there are other ways to approach joint exploratory activities that need not involve such intimate melding of hardware.

One potentially efficient approach could be based on the practical realization that a major portion of the cost of a manned space project is devoted to reliability, back-up systems, contingencies, planning for emergencies, etc. A 90 percent reliable system may cost $X million, a 95 percent reliable system $2X million, and a 99 percent reliable system several times as much. Today's shuttle astronauts probably spend half their time practicing maneuvers they will never have to perform in space (such as aborts), and a great deal of the rest of their time is spent on familiarization with contingency procedures. That, too, is very expensive—but vital.

If both the U.S. and USSR were operating manned vehicles at Mars (as, for example, they are today in Antarctica), the very existence of such systems purely for emergency rescue purposes should allow significant cost reductions for both nations. The reason for this is that a lower single-spacecraft reliability would be acceptable if another spacecraft could be counted on for "worst case scenarios." Even a slight reduction in the required reliability percentage could cut the total program cost in half, or better.

For example, if the Soviets had a base on Phobos and could be counted on to fulfill their current treaty obligations regarding the rescue of endangered spacefarers, an American spacecraft orbiting Mars could do without some of the third and fourth backup systems and without the excruciatingly expensive equipment and the procedures qualification activities that would be needed if the astronauts were entirely on their own. The "do-or-die" requirements of their earth-return rocket engine could also be significantly relaxed, resulting in additional major cost savings.

It should not be forgotten that we are talking about base costs of manned Mars orbital operations on the order of half or less of one Apollo, the normalized unit cost of the man-to-the-moon program of the 1960s. So even without savings, the expeditions are already affordable by a single country. Cost savings by splitting up the hardware may be mostly illusory to begin with and, as mentioned, a political dead end as well. Cost savings based on parallel independent operations, on the other hand, promise to be substantial—without the threat of mutual veto.

Such a strategy is also stable in the scientific sense, that is to say, deviations from the partnership will generate "restoring forces" rather than the tensions, fears, and mistrust that could uncontrollably tear apart a mixed-mode cooperation strategy at the first sign of diplomatic difficulties. The reason for this is that without one side's participation, the other side could proceed with its own homegrown hardware, but at a recognizably higher risk to the lives of the spacefarers. In the real world, a Soviet unilateral pullout would make for immensely bad public relations, since it would merely succeed in increasing the danger to the American astronauts without stopping them cold. If in a worst case the absence of a promised Soviet "safe haven" rescue option did actually result in fatalities, the damage to the USSR's international prestige would be literally cosmic in scale. Such a possibility, therefore, should produce a large inducement for them to stay in the agreement despite any temporary bilateral diplomatic problems; such an inducement does not exist in the mixed-mode strategy, which actually may provide all-too-tempting blackmail opportunities for late-inning threats of pullout or delay.

In this analysis, what I call the parallel joint strategy is thus inherently stable, while the current favorite, the mixed joint strategy, is inherently unstable. Neither has any clearcut cost advantage, although if forced to make an intuitive judgment, I would opt for the parallel strategy as the dollar-wise cheaper alternative.

We can also cooperate with the Soviets without having to like them, despite the American urge to think of all our teammates as "the good guys." We were co-belligerents against a greater danger in 1941–1945 but, if anything, Stalin ran an "evil empire" worse than Hitler's—except for the moment he was not bent on world domination. We can cooperate today in other fields of common need, but we don't have to forget the genocide going on in Afghanistan, the spiritual crushing of Poland, the brutal destruction of an innocently lost airliner and the horrible deaths of the 269 people aboard it, the mountains of lies and hypocrisies that strangle the souls of ordinary Soviet citizens. For reasons, we can cooperate with the USSR, or South Africa, or even daffy Qaddaffi—for good reasons.

Space may be such a reason, and may be worth the moral association. But if so, we should forge the alliance with no illusions. And most of all, I urge that we design the team so as not to give known enemies the power and the temptation to unilaterally pull the plug on our future.

Ad astra per aspera! Who ever promised a rose garden, on earth or off it?

18

In Defense of Manned Space Flight

S. FRED SINGER

Most of the public takes manned space flight pretty much for granted, including human exploration of the moon and planets. Many accept the idea of eventual settlements on extraterrestial bodies, perhaps with mining ventures to extract useful resources. Some would even set up human colonies in interplanetary space.

Such views of the future are not accepted by a substantial number of experts. While a minority within the public at large, they may well form a majority within the space science community. One of the most outspoken of these scientists has been Prof. James A. Van Allen, with whom I was closely associated during the early years of the "space age." We lofted scientific instruments above the earth's atmosphere in V-2 rockets that had been captured during World War II—long before there were scientific satellites.

It would therefore be quite natural for me to sympathize with Van Allen. Certainly, instruments perform functions that no human observer could tackle, generally at lower cost and without risk to life. Yet I disagree strongly with his conclusions, which would effectively terminate the manned space program.

Here is what Van Allen has to say, in a recent issue of *Science* (May 30, 1986), the journal of the American Association for the Advancement of Science:

> Closely akin to science fiction and a prominent part of the 1986 scene is the large number of futuristic proposals for space fight. I may mention a few by short title: solar power satellites; manufacturing in space; permanently manned space stations in earth orbit, on the moon, and on Mars; and the economic mining of asteroids.
>
> It is difficult to distinguish the proposals of prophets from those of

198

charlatans, and I am not so foolish as to suggest that such undertakings are out of the question at some remote time in the future. But not one of them can withstand critical scrutiny in the context of the present century. I consider that untimely advocacy of them, especially by prominent national figures, does the entire space effort a disservice.

He correctly points to the past achievements of instrumented, unmanned spacecraft:

> Many space enthusiasts blithely ignore the fact that almost all the truly important utilitarian and scientific achievements of our space program have been made by instrumented, unmanned spacecraft controlled remotely by radio command from stations on the earth.

As an early proponent and designer of instrumented earth satellites, I must agree with this statement. But I do not accept much of what follows:

> Despite the great successes and future promise of automated, commandable, unmanned spacecraft in providing vital human services and scientific advances, the President and Congress persist in giving primary emphasis to the misty-eyed concept that the manifest destiny of mankind is to live and work in space. The proposed development of a system of permanently manned space stations serves this concept but is otherwise poorly founded. . . .
>
> Fervent advocates of the view that it is mankind's manifest destiny to populate space inflict a plethora of false analogies on anyone who contests this belief. At the mere mention of the name of Christopher Columbus they expect the opposition to wither and slink away. . . .
>
> But the application of the Columbus analogy to support advocacy of a manned mission to Mars is massively deceitful. Mars is not terra incognita. We have already explored it and found it to be far more desolate and sterile than the heart of the Sahara desert. There, of course, remain many matters of deep scientific interest on Mars but these matters can be addressed systematically—at much less cost and without risk to human life—by automated, commandable spacecraft, surface rovers, and sample return-to-earth missions.

But Mars has hardly been explored. By analogy, imagine two Viking landers, with their primitive instrumentation, drawing conclusions about life on the planet earth after descending onto an Antarctic ice plain! Mars is a very complex planet whose geologic and climatic history can tell us much about the earth.

The evidence suggests that volcanism on Mars extended over most of its existence, that the climate of early Mars was wet and warm, and that large amounts of water may still be locked up in Martian permafrost.[1] The discovery of crypto-life, or even fossil life, on Mars could affect biology in a fundamental way and would change the probability of the existence of extraterrestrial life from near zero to near one.

I also take issue with the assertion that Martian scientific exploration can be conducted at "much less cost . . . by automated, commandable spacecraft, surface rovers, and sample return-to-earth missions." I have studied this matter in some detail and concluded that the least-cost solution involves, after unmanned precursors, a manned presence on the Martian satellites, Deimos or Phobos.[2] Reaching them is relatively easy, requiring little propulsion, aided by aerobraking in the Martian atmosphere.[3] The two to four months' operation would direct, *in real time*, some dozens of *un*manned rovers, one at a time, to different Martian locations to bring back surface and subsurface samples to the Deimos-based laboratory for detailed examination. This *sequential* experimentation permits efficient exploration of the most promising and often most difficult terrain for a study of the geologic, climatic, and possibly biologic history of Mars. A few rovers can be lost, and will be lost, without jeopardizing the mission.

By contrast, directing an equivalent operation from Houston runs into the fundamental limitation of signal travel time, a problem that can be ameliorated but not solved by "smart" rovers. Data rates would be much lower; returned samples, containing volatiles, would no longer be fresh; contamination may become a problem. A complete study of Mars, based on sequential landings of rovers, would have to extend over decades and be more costly and scientifically inferior to a manned operation involving the Martian satellites.

Van Allen's proposed policies would spell the end of manned space activities. His two leading recommendations are:

- Suspend manned flight indefinitely pending critical assessment of its justification
- Postpone development of the space station

while he proposes an expansion of the unmanned space science program.

But manned spaceflight is much more than just space science; that's too narrow a view. There is the study of human physiology under space conditions; the task of devising, adjusting, servicing, repairing, and retrieving (if necessary) multi-billion-dollar scientific payloads; setting up and managing space manufacturing activities; possible military applications. Other considerations may be Soviet competition; national prestige; and just plain adventure, with tourism probably not too far away.

In any case, it is unrealistic to imagine that the public will continue to support expensive space science projects without a manned component. There is, after all, other important science in need of funds that is of more fundamental interest or addresses basic human needs: biology, medicine, high-energy physics, even ground-based astronomy, to mention but a few.

I started my career as a proponent of instrumentation and opponent of manned space operations, considering them to be little more than publicity stunts without much scientific justification. My "conversion" to a broader viewpoint dates to less than twenty-five years ago. After having directed the development of an operational weather satellite system I began to realize that manned operations could be cost-effective in the development of advanced and more costly systems. I still believe that there are specific projects, perhaps not too many, where a manned presence can produce an overall improvement, at lower cost.

NOTES

1. R. A. Kerr, Science 233, 939 (1986).

2. S. F. Singer, "The Ph-D Proposal: A Manned Mission to Phobos and Deimos" in *The Case for Mars*, ed. P. Boston, Amer. Astronaut. Soc., San Diego, 1984, based on a study conducted for NASA during 1977–78. The earliest exposition of the Ph-D proposal appears in *Explorer*, 15, no 2 (Jan. 1978), Am. Inst. Aeron. Astronaut., Huntsville, AL. For an even earlier discussion see S. F. Singer, "Manned Flight to the Nearer Planets," in *Manned Laboratories in Space*, ed. S. F. Singer, 1968 Int'l. Acad. Astronaut. Symp., Reidel, Dordrecht, 1969.

3. J. R. French, "Aerobraking and Aerocapture for Mars Missions" in *The Case for Mars*.

PART VIII

SEARCH FOR EXTRATERRESTRIAL LIFE

19

The Search for
Extraterrestrial Life

G. SETH SHOSTAK

Introduction

The possible existence of extraterrestrial life has, since the second World War, become a matter of serious inquiry. This is largely due to the sudden expansion of knowledge in both biology and astronomy, which has provided new a priori reasons to expect that life is common in the universe. But it is also the result of advances in space technology and electronics that have finally made it possible to seek out extraterrestrials with a real chance of success. The search for life, often designated by an acronym of more limited scope, SETI (Search for Extraterrestrial Intelligence), has moved from the province of science fiction to that of science endeavor.

From a philosophical point of view, the existence of intelligent aliens has seldom been in doubt. The more enlightened thinkers of classical times assumed that heavenly bodies were host to other beings, and not only the residences of the gods. But early Christianity disputed such ideas, basing its opposition not only on theological grounds, but also on the Ptolemaic earth-centered astronomy.

The Copernican revolution, which expelled man from his central position in the universe, also saw a change in the popular view regarding cosmic habitation, and this was largely sanctioned by the eventual conversion of the Church to the new astronomical order. When the crude telescopes of the seventeenth century showed the planets to be other worlds, the temptation was irresistible to assume they were, like the earth, inhabited by intelligent beings. The moon, and even the sun, were soon populated in both public and scientific opinion. But as optical instruments improved and astronomy became more exact, the advocates of extraterrestrial life began to focus more and more of their attention on Mars. Nearby and, unlike Venus, possessed of a transparent atmosphere, the surface of Mars as viewed through the telescopes of the last hundred years provided several hopeful indications for life.

A frequently reported phenomenon was that the dark areas of Mars were green, and this was readily interpreted as due to widespread vegetation. Unfortunately, there was little truth to either the reports or the interpretation. The chromatic aberrations of the older refractive telescopes tended to spread bluish-green light throughout an image focused in the red (the dominant color of the brightly illuminated Martian surface). In addition, one must remember the eye's propensity of filling in dark areas of vision with the complementary color of bright areas.

Of greater fame than the Martian vegetation was the claim that the Martian desert was laced by long, straight canals. To some extent this was the result of semantic confusion: Schiaparelli's description of such features a century ago was badly translated from the Italian ("canali," meaning channels, became "canals," which are deliberate constructions). Nonetheless, the American astronomer, Percival Lowell, made convincing and elaborate arguments for the existence of an extensive network of canals in the early part of the present century. Once again, the eye seems to have played tricks on the observer by tending to connect scattered surface detail into linear structures. Thanks to present-day space exploration, we now know that the claimed canals were a fiction although, ironically, long, linear, and naturally formed canyons do exist on Mars.

Despite these historical misconceptions, until quite recently Mars continued to be the principal hope for company in our Solar System. Possessed of an (albeit thin) atmosphere, a hard surface, and moderate temperatures, Mars could conceivably harbor simple plant life, even if there was no evidence for advanced beings. Unfortunately for this view, the Viking landers that performed elaborate and sensitive tests for biological activity on the Martian surface failed to detect anything suggestive of life. The negative results of these probes could be questioned by the skeptical, but to paraphrase one researcher: it's possible that the experiments did indicate life, but it's also possible that the rocks seen at landing sites are actually living organisms that look like rocks.[1] Despite centuries of speculative literature, and an abundance of suggestive cinema, Mars today appears distressingly dead.

Unlikely as it may seem, Jupiter might be the last hope for extraterrestrial life in the Solar System. It's conceivable, although improbable, that deep in the 1000 km-thick atmosphere of this planet, a sort of microbal life floats in the churning mists. There, in a never-never land between the Jovian cloudtops above and a gloomy, bottomless, black sea of liquid hydrogen below, these postulated suspended life forms could enjoy temperatures comparable to those on earth's surface.

However, despite this and other interesting possibilities, the conclusion most researchers draw from three decades of space exploration is that we are alone in the Solar System, a disappointing culmination to thousands of years of optimistic expectation. Nonetheless, SETI programs today enjoy considerable support; in 1982 a special commission of the International Astronomical Union was set up to deal with exobiology and SETI. The apparent contradiction between past result and current initiative is easily explained: we have failed to find life in our own backyard, but we are gaining the means to extend the search to other stars. How should we do so, and where should we look?

The Modern Astronomical Facts, or Should We Bother to Search?

1. How Many Stars?

The Copernican revolution, which displaced man from his preferred position in the universe, continued well into this century. A hundred years ago, all stars were thought to be part of a massive single, flattened system, the Milky Way galaxy. Our own sun, together with its attendant planets, was believed to be near the center.

In the 1920s, astronomers were finally convinced that many of the faint nebulae visible in large telescopes were in fact galaxies in their own right. Furthermore, Kapteyn and other early investigators of the Milky Way had neglected to consider the effects of interstellar dust on their star counts, and this had misled them into concluding that the sun is near the center of the galaxy. It is now known that we are situated in an unremarkable fringe area of a rather unremarkable galaxy. This in itself is a powerful philosophical argument against the view that life is unique to earth, or to our Solar System.

The sun is not only rather indifferently located, but is also of a common sort. The total stellar population of the Milky Way is approximately 100 billion stars, of which perhaps one in a hundred is similar in size and temperature to the sun. In our own galaxy, the sun has a billion brothers.

2. How Many Planets?

How many of these suns have planets? The detection of solar systems beyond our own is a difficult observational problem and, as discussed below, the classic evidence based on small perturbations in the motions of stars is inconclusive. In the recent past, the slow rotation of late-type stars (like the sun) was thought to be convincing evidence that planets were a general phenomenon. The argument ran as follows: normally, as stars are born out of a collapsing gas cloud, one would expect them to increase their spin, as the cloud shrinks. But if they could somehow transfer their angular momentum to an encircling planetary assemblage, then this would explain the observed lack of fast rotation (note that 99 percent of the angular momentum of our own Solar System resides in the planets).

But the fact that stars similar to the sun spin slowly has found a new explanation in the recent discovery that mass loss, the ejection of stellar material, is a common phenomenon even among late-type stars; apparently this process alone could account for the slow stellar spins.

While the long-standing indirect arguments for planets have waned, they have been more than compensated for during the last year by a burst of observational evidence for other solar systems, all of it from new telescopes or new methods of using conventional telescopes. One of the most exciting discoveries of the IRAS infrared satellite was the existence of large rings of material around Vega and several dozen other stars.[2] The particles comprising the rings are at least a few millimeters in size, or ten thousand times larger than the average interstellar dust particle. Chunks of material, up to and including planet size, could also be present, but the IRAS observations would not necessarily detect them. Despite this uncertainty, the existence of extended disks of cold material encircling a star gives encouraging support to scenarios in which planetary systems are a natural accompaniment of star birth.

Tipped off by the IRAS discoveries of a year ago, visiting astronomers in Chile attached a special CCD ("charge-coupled device") imaging detector to the 2.5 m diameter telescope at Las Campanas. CCD detectors are many times more sensitive than film, and allow very faint light to be recorded. Furthermore, the fact that CCDs have a large dynamic range allowed the observers to carefully mask off light from the star they were measuring, and even get rid of telescope defects by subtracting the response of another star of similar brightness. The two astronomers found an apparent ring of material around the star Beta Pictoris, extending about 15 arcseconds from the star.[3] (Beta Pictoris is a relatively nearby star, at a distance of 50 light-years.) This is very direct evidence for a protoplanetary disk.

The latest news in the race to find planets is the claim by astronomers at the University of Arizona to have discovered a massive companion around the star Biesbroeck 8, a nearby star about 21 light years distant.[4] The astronomers made use of the technique of speckle interferometry to cancel out much of the blurring effect of the earth's atmosphere, thereby revealing a sphere thirty to eighty times the size of Jupiter orbiting Biesbroeck 8. The surface temperature of this giant gas ball is 1,100° C, and the question remains whether one should refer to such an object as a planet, or a "brown dwarf," a kind of low temperature star. This may be more a question of semantics than substance, however. (In a sense Jupiter is a "star" since it emits more energy than it receives from the sun.)

In summary, then, while indisputable evidence for extra-solar system planets is still wanting, the last year has seen direct observations of material, most likely protoplanetary, about several nearby stars. These observations give substantial impetus to the point of view that planetary systems are a quite natural adjunct to star formation. Planets may be the rule, and not the exception.

3. *The Bottom Line: How Many Earths?*

How many planets are capable of supporting life, that is, are favored with moderate temperatures, suitable surface gravities, and atmospheres? The statistics of our own Solar System suggest one in nine, but even if the number is one in a thousand, then ten million earth-like planets orbit solar-type stars in the Milky Way. We have assumed that all late-type stars have ten planets, but to be conservative imagine that only 1 percent do (such a low percentage will also take into account the fact that about half of all stars are in multiple systems, with their resulting unstable planetary orbits). Then the number of sister "earths" declines to a measly hundred thousand.

At this point the reader may object to the summary manner in which we have attempted to estimate the number of planets in our galaxy. Unfortunately, a better estimate is presently impossible, since the only hard astronomical fact is the number of solar-type stars in the Milky Way. But the bottom line is that we are dealing with large numbers, and uncertainties of a factor of two, of ten, a hundred, or even a thousand, do not qualitatively alter the conclusion that there must be many, many planets capable of supporting life in our galaxy. And we remind the reader that the Milky Way is but one of millions of galaxies.

If "earths" are plentiful, then is life also? It is not the purpose of this paper to discuss how life first begins. We note that many (but certainly not all) biologists favor the view that, given the requisite environment, life will ultimately cook out of the primordial "soup" that washes the shores of a young planet. As possible support for this contention, we remark that life started on earth virtually as soon as the oceans had formed. The earliest known microfossils are at least 3.5 billion years old.[5] In other words, as soon as life could begin, it did begin. If this is the typical situation, then our sister "earths" can also be expected to develop life. Whether this life invariably evolves to intelligent life is, again, a question beyond the province of this paper; many researchers have pointed to the excruciatingly slow, and seemingly accidental, evolution of hominids on earth.[6] At the moment, we have no choice but to assume that the one example of evolution we have is typical, and that the ultimate production of intelligence is not wildly improbable.

We have rather quickly traversed a chain of reasoning and assumptions that suggests that life may be a common phenomenon in our galaxy. The uncertainties are large, but so is the number of stars. On balance it seems worth the effort to search beyond the Solar System. How should we do so?

How to Find Them

1. Active Strategies

Direct contact with alien civilizations by simply blasting off to other stars in rockets is a scheme perennially popular in fiction, and intermittently popular with scientists. Our present-day spacecraft are capable of leaving the Solar System with velocities of a few dozen miles per second. At that speed, the nearest stars are at least a 50,000-year ride away, and quite obviously unattainable.

Nuclear-powered rockets could conceivably shorten this travel time by a factor of a thousand, and a matter-antimatter energized craft could be yet another factor ten faster, but the technical problems involved in either of these approaches are very far beyond our forseeable capabilities. We cannot yet go to the stars, although as discussed below, the aliens may be able to do so.

While interstellar physical exploration is presently not feasible, communication by radio might be. It is already the case that leakage radiation from early radio and television broadcasts has reached hundreds of nearby stars. Should these signals be intercepted, and a reply sent in our direction, one could speak of the establishment of communication. Unfortunately, the turnaround time for messages even to the nearest stars is measured in decades, so despite occasional broadcasts into space designed to increase interest in SETI, most actual SETI effort is expended in passive projects: listening via radio telescopes for transmissions either intentionally beamed our way, or susceptible to chance interception.

2. Passive Techniques, or Playing Eavesdropper

a) Radio

The bulk of all serious attempts to detect life beyond the Solar System have involved listening at microwave frequencies for artificially produced signals, using radio antennas built for astronomical research. The rationale for doing so is that electromagnetic radiation represents the fastest and least expensive (in terms of energy) carrier of information across interstellar space.

While the universe is transparent to most electromagnetic radiation, quantum noise and Galactic non-thermal (synchrotron) radiation dominate at high and low frequencies. When these unavoidable noise sources are summed, we find a minimum, often called the "free-space microwave window" between frequencies of 1 and 60 giga Hertz (GHz).* Our own atmosphere (and presumably that of the aliens, as well) has absorption lines of oxygen and water above 10 GHz, which compels us to consider the low frequency end of the window. Here lie the natural emission lines of neutral hydrogen H

---

*1 billion cycles per second; a wave length of 30 cm.

(1420 MHz) and the hydroxyl radical OH (1662 MHz), defining a small spectral region referred to as the "water hole."

Deliberate modern radio searches total a few dozen, and date from the celebrated Ozma project of 1960. Frequencies observed range from 0.5 to 22 GHz, but most are in the "water hole," with the favorite frequency being the 21-cm H line, at 1420 MHz. The majority of the observations were made either in the U.S. or the USSR, and most have concentrated on nearby solar-type stars. Until now, no confirmed detections of non-natural signals have been made.

For radio searches of the future, the primary parameters remain (1) which direction to look, (2) at what frequency, and (3) with what frequency and temporal resolution. Questions such as how large should our antennas be are, for simple financial reasons, not terribly relevant: SETI will be conducted with the radio telescopes presently available.

Nonetheless, it is an obvious desideratum of any search that we make allowances for civilizations that might be quite different from our own.

As an example of current thinking, we remark on the search strategy envisioned by the American SETI Science Working Group (SSWG) for the next decade.[7] The strategy is two-pronged, consisting of a relatively intensive study of nearby, and likely, stellar systems, and a much lower sensitivity search of the "whole sky." (In practice, the search would be concentrated toward the Galactic plane). The former is intended to be able to detect the leakage radio emission from a civilization much like our own, whereas the latter is intended to pick up intentional, powerful radio beacons broadcast by advanced societies.

Probably not coincidentally, this two-sided strategy is a good match to presently available radio astronomy antennas. Sensitive searches are best done on the largest available telescopes, but these have only limited observing time that can be devoted to SETI. The all-sky search, which by definition is of considerably lower sensitivity, can be carried out by smaller antennas, and in some cases these can be 100 percent dedicated to SETI projects. One may remark that such a pragmatic approach is qualitatively different from the ambitious CYCLOPS plan of a decade ago, in which it was hoped to build an extensive, two-dimensional array of moderately large antennas, exclusively intended for SETI work.[8] No one would deny that such an instrument would promise a greater chance of success, but the new plan is more in tune with present economic realities.

The details of the SSWG plans are based on our current feelings about alien engineering. The most economical means (in terms of energy) for transmitting information is to use coherent signals, that is, narrow in bandwidth and/or time. We assume that the aliens are also energy-conscious, and that their signals will be coherent in this engineering sense. There is an important benefit to such an assumption, namely that it provides a simple way of discriminating alien signals from the "natural" signals of space. The latter are almost always incoherent (two exceptions being pulsars, exhibiting sharp temporal structure, but broadband frequency spread, and interstellar masers, narrow in frequency but without regular temporal variation). In other words, we will be able to recognize the aliens because their signals will have narrow frequency and/or temporal spectra, which can easily be sifted out from the astrophysically-generated background noise.

Dispersion by interstellar gas limits signals traversing space to a minimum bandwidth of about 0.01 Hz. But to actually observe with a filter of this width, while at the same time covering an interestingly broad segment of the spectrum, would require an unreasonably large multichannel receiver. Consequently, actual observing bandwidths of 1 Hz have been proposed for the intensive searches of nearby solar-type (and perhaps other) stars. For the "all sky" search, a bandwidth of 32 Hz is proposed. These bandwidths are similar to those of many terrestrial signals.

It is worth remarking that previous radio searches have typically used bandwidths measured in kilohertz. That is because these are the resolutions that are optimal for radio astronomical research. Certainly one of the greatest incentives for further SETI work with existing telescopes is the rapid pace of progress in digital electronics, which now allows the building of inexpensive autocorrelator receivers having many channels and providing the narrow bandwidths appropriate to searching for artificial signals.

How sensitive could a present-day search be? According to the SSWG report, the 1000-ft diameter Arecibo telescope could detect a transmitter of 100 Megawatts effective radiated power (ERP) at the distance of the nearest stars with a ten-minute observation time. The "all sky" survey will use smaller telescopes, and will observe each position for only a few seconds at most; consequently its sensitivity is several hundred times poorer. For comparison, the strongest terrestrial TV transmitters have powers of 0.1 to 10 Megawatts, and the most powerful radar systems 10,000 Gigawatts (both ERP). Thus, while we may be deprived of receiving the aliens' television entertainment, we could detect their missile defense systems at more than a thousand light years distance.

b) Optical

Radio as a means of communication continues to get most of the attention of the SETI community, but there are other ways to search out extraterrestrials. Radio is an advanced technology on earth, but only in the last twenty-five years have we begun to develop the possibilities of communicating at optical frequencies, using lasers. As anyone aware of the current interest in glass fibers knows, light as a signal carrier offers the big advantage of very wide bandwidth, that is, large channel capacity. A single beam of light could easily carry all the world's telephone traffic.

How feasible is it to signal between the stars using light? Lasers are coherent light sources; that means that they can be sharply focused, or beamed. A focused ray is desirable for maximizing the reception strength when signaling in the direction of a particular star. A ten-inch lens set in front of a laser could produce a one-half arcsecond beam, or narrower than the largest radio telescope arrays could manage at 21-cm wavelength. And the transmitter power achieved by pulsed lasers is already approaching that of large radio transmitters. On the face of it, then, optical transmitters appear to be smaller, and of much larger bandwidth, than their radio counterparts.

Nonetheless, there are fundamental difficulties with interstellar optical communication. To begin with, quantum noise in any receiver increases at high frequencies. The intrinsic noise power in a radio, given by kT,* is fixed by the equivalent noise temperature of the receiver, T. At optical frequencies, however, the noise power density goes as hf.* The consequence is that, despite the much sharper beaming of a laser, a radio transmitting antenna twenty inches in diameter will occasion the same received signal-to-noise as an optical mirror ten inches in size, given identical transmitter powers. And since it is easier to build very large, accurate radio antennas than to construct very large, accurate mirrors (and the latter would only make sense in space, where there is no atmosphere to defocus the light beam), radio may remain a preferred interstellar communication technology on practical grounds.

Aside from the signal-to-noise difficulties caused by the equipment, there is also the problem of natural radiation sources swamping the signal. At typical stellar distances, a light signal beamed from a planet would appear to be separated from the planet's sun by only a fraction of a second of arc. One strategy for ameliorating the resulting confusion would be to transmit in spectral bands where one's sun would be dimmer, that is, infrared or ultraviolet. This may place a burden on listeners, however, since their own atmosphere may block the signal before it reaches their telescopes. Another idea would be to transmit at those few special wavelengths that coincide with strong stellar absorption lines (for example, the H and K lines of calcium, or the well-known yellow sodium D lines). At the bottom of such absorption troughs, the star's radiation would be an order of magnitude weaker than at adjacent wavelengths, and listeners could spectrally separate the transmitted signal from the natural emission of the star.

At the moment, large terrestrial radio telescopes (such as the 100-ft diameter Arecibo antenna) could, assuming comparable installations on other worlds, communicate at distances of thousands of light years. Our capability in the optical region is at least two orders of magnitude poorer, and may continue to remain inferior to radio for the reasons given above. Nonetheless, it would be presumptuous to

*k is Boltzmann constant; h is Planck's constant; f is frequency of radio signal.

rule out a technology in which our experience is still so limited. The ironic fact of SETI experimentation is that we will know the optimal search strategy only after the experiment has succeeded.

c) Optical Discovery Techniques

Aside from communication by optical means, there are several ways in which optical observation may help us to discover extraterrestrial life, even of a non-intelligent sort,—or at least help us to better estimate the chances of life. To begin with, the discovery of other solar systems is a project based exclusively on optical observations. Historically, this has been perused by looking for the "wobble" that a star accompanied by a massive planet undergoes as they waltz about their common center of gravity. This involves comparing high precision photographs of the candidate star taken decades apart. There have been numerous claims of planetary systems discovered by this classic, astrometric technique in the past, but a contemporary appreciation of the systematic errors inherent in the measurements has thrown them all into doubt.

But even if no fully reliable detection of planets around other stars has yet been made, improved astrometric techniques and the advent of the Space Telescope will provide an opportunity to change this situation. As discussed above, the infrared satellite IRAS, as well as improved methods of direct telescopic observation, have provided dozens of new candidates for planetary or proto-planetary systems, and these will be closely investigated by modern astrometers. Of course, astrometry projects begun now take decades to supply answers.

Another possibility is to directly observe the orbital motion of a star about a displaced center of gravity. Our sun, for example, circles about the sun + Jupiter barycenter with a velocity of roughly thirteen meters per second. A high-dispersion spectroscope capable of detecting such small velocity perturbations in other stars is probably feasible in the near future. Spectroscopy can furnish answers quickly but, like the astrometric technique, is limited to detecting only larger planets: the orbital motion of the sun due to the earth is only 0.1 meters per second.

In summary, contemporary optical astronomical instruments provide several good possibilities for detecting other solar systems. Since this is a classic research problem, it will be pursued irrespective of the vagaries of SETI funding. Its importance to the subject at hand is obvious since we have assumed that life can develop only on planets.

While not of immediate concern or application, it is worth noting that, should we succeed in finding other planets, and if we could somehow measure their optical spectra, then life might reveal itself in the planet's atmospheric composition.[9] Oxygen, for example, is abundant in the air only because of the biomass. A nice advantage of this technique is that it is capable of detecting life that isn't intelligent.

We Haven't Found Anything. What Does That Mean?

We have seen that a quarter-century of spaceflight has driven us to the conclusion that the rest of the Solar System is probably lifeless. And despite a similar period of SETI enterprise, no confirmed alien signal from the stars has been detected either.

At first glance, there is no reason to lose hope. For passive searches the parameter space is large: frequency, polarization, direction, power, time resolution . . . these are all unknown quantities, most of which have an enormous range of possible values. Frank Drake has referred to this as a "cosmic haystack," and our search for the needle has so far entailed only one part in 10^{17} of the hay. New receiver technology will increase our search space by orders of magnitude in the near future and, of course, we have assumed that there are many needles in the stack.

Nonetheless, our failure to hear the aliens may imply something profound. Earlier in this paper we made a very rough estimate that the number of "earths" floating about the Milky Way is one hundred thousand. As our own solar system is relatively young—4 billion years old against 15 billion years for the age of the galaxy—most of these other earths have had billions of extra years to develop life. The majority have had the opportunity to spawn intelligent species.

Now, any advanced civilization will invent medicine, and as a result will eventually fill its planet with a burgeoning populace. Orbiting space stations, such as proposed by O'Neill, will be built.[10] Contrary to one's likely first reaction, it's probable that life in these artificial environments can be made quite pleasant, and new generations of beings will think nothing of being born, raised, and dying in space.

With little additional effort, these space stations can be propelled to nearby stars. Depending on the technology used, the interstellar travel time might be hundreds, or even thousands, of years. But that's O.K. After reaching a habitable system, the "colonists" might take a thousand-year break, to establish their culture and increase their population. Then they, too, will send off some colonists, who will spend a millennium getting to the next star. And so forth. The colony will slowly spread, as Frank Drake has said, like coral in the sea.

Now the point is, they will run out of new stars in 10 million years, or less than one-thousandth the lifetime of the galaxy. In a blink of an eye, relatively speaking, the entire Milky Way could be colonized. One might liken the phenomenon to the expansion of the Spanish Empire in the sixteenth century: once Columbus made his first voyage, the whole Caribbean and Pacific were visited in but a matter of decades. There were Spaniards everywhere.

Therefore, the aliens should have been everywhere, including our own Solar System.

Now, the followers of Von Däniken believe they have been here, and the more optimistic members of the UFO crowd think they still are. But the hard-nosed, irrefutable evidence is missing. Fact A of the SETI business is that we seem to be alone, and we shouldn't be.

Enrico Fermi summarized the problem in one sentence: "Where are they?"

Could We, Then, Really Be Alone?

The argument that the aliens should have completely colonized the galaxy is, in fact, a very strong one. Many of the parameters, such as travel times, number of habitable planets, etc., are uncertain, but varying these within an order of magnitude or two doesn't change the conclusion.

The most obvious solution to this paradox is that we really are alone. Hart has pointed out the complexity of the nucleotides that make up DNA; he figures that the random combining of atoms in the primordial oceans of a distant earth will take 10^{32} billion years before dishing up something interesting.[11] In his view, not only are we alone in the Milky Way, but probably in all the universe we can see. Fred Hoyle has put it another way: the chance of forming life in this manner is the same as assembling a Rolls Royce by sending a tornado through a junkyard.

Nonetheless, the idea that we are astronomically very ordinary, and yet somehow biologically unique, smacks strongly of the anthropocentrism that used to characterize our view of the heavens. Less radical ideas have been put forth to explain our apparent solitude. For example, it has been suggested that technological civilizations inevitably, and rather quickly, self-destruct, before getting the chance to embark on space colonization. Or that the only societies that avoid this fate are those that develop passive lifestyles, with philosophies incompatible with searching for new worlds. Either of these scenarios would, of course, explain our failure to detect extraterrestrial signals.

Drake has submitted that the costs of interstellar travel, no matter what technology is used, are simply too high in terms of energy.[12] This could explain the lack of aliens in the neighborhood, but not the absence of signals. But then again, perhaps nondirected radio is a very temporary technology: soon we can expect all terrestrial broadcasting to occur via non-leaking glass fibers and beamed microwave and optical links. If the aliens also go this route, it might be quite difficult to eavesdrop.

Nonetheless, as reasonable as such explanations might be, it's hard to imagine that they are relevant to every single society in the galaxy. Remember: only one civilization has to be adventurous, and even the most passive society would certainly feel compelled to develop interstellar travel if its existence were being threatened by an impending supernova of its home star.

Is It Dangerous to Signal?

The lack of detected signals prompts one to ask whether it is prudent for any civilization, including ours, to casually disclose its existence. Carl Sagan has not hesitated to send maps of our where-abouts into space by gluing them onto spacecraft (admittedly such slow-moving calling cards have a low probability of ever being re-ceived). He has also used the Arecibo antenna to beam radio signals to a globular star cluster, although even this speedy form of com-munication cannot be expected to elicit a reply for a half-thousand years.

Should Dr. Sagan be praised for his inventiveness in trying to signal the aliens, or is he destined for eternal obloquy as the man responsible for the future obliteration of earth at the hand of ag-gressive extraterrestrials, lured here by his messages? Most of the popular notions regarding the evil intentions of aliens can be dis-missed as the work of science fiction writers looking for conflict. And for practical reasons, extraterrestrials are unlikely to be very inter-ested in destroying our civilization simply to gain access to our Solar System's natural resources (the latter does not require the former, and besides, transport costs probably rule out mining someone else's solar system). Nonetheless, even Sagan himself, in his book with Shklovskii *Intelligent Life in the Universe,* remarks on some of the irrational treatment of American Indians by European explorers of the sixteenth century, and comments on the possibility that the aliens could be motivated by a sort of extraterrestrial evangelism: the con-version of other worlds by peaceful means if possible, but by force if necessary.

Such arguments have bearing on Von Hoerner's suggestion that only passive societies survive in the long run.[13] A civilization without aggressive impulses is not only less likely to be motivated to beam radio beacons into space, it may perceive danger in doing so.

Are We Like Ants Trying to Communicate with Dogs? Or with Robots?

Most of the scenarios for contact described here assign us a passive role: we listen for messages beamed our way, or search for interstellar probes sent by other societies. Bracewell has speculated that in the 15-billion-year-long history of the galaxy, contact among civilizations will have happened often, continually bringing emerg-ing, new civilizations into a sort of "Galactic Club."[14]

While this is an appealing idea, there may be reason to be less optimistic about such congenial behavior among the galaxy's inhabi-tants. Bracewell himself has remarked on the fact that the explorer always finds a civilization less advanced than his own. Columbus came from a society of greater technical sophistication than the Indian society that he discovered; otherwise the Indians would have landed in Spain, rather than the other way around. Our society has just reached a technical level that allows the possibility of contact with the aliens, so it is self-evident that if we do establish communi-cation, it will be with an advanced (and probably very advanced) society. If our own behavior is any guide, it seems questionable whether such a society would be interested in having the clearly inferior earthlings join their "club."

We don't invite pigs or porpoises into our homes to watch television, and these creatures may be rather closer to our intellectual level than the extraterrestrials whose radio beacons we might intercept.

Indeed, the science-fiction idea of a galactic "empire" might be more realistic. As discussed above, the number of planets in the galaxy that now have, or ever had, intelligent life could be any number between one (our own) and a billion. But if this number is relatively small, say a thousand, then biological evolution will have ensured an interesting diversity.

Imagine that the start of life on the thousand planets that house these civilizations occurred randomly during the course of the galaxy's history. Then the minimum average difference in evolutionary progress on these planets (assuming, for simplicity, that they all evolve at the same rate) would be 15 million years. Even this minimum difference is substantial in evolutionary terms; man descended from the apes in a fifth the time. The conclusion one draws from this is that the first civilization capable of colonizing the galaxy would be able to do so without any effective competition.

In other words, if the occurrence of life is infrequent, and if societies capable of interstellar travel are inclined to undertake it, then mastery of the galaxy would engender far less resistance than the Spaniards encountered in their conquest of the Americas. The galaxy might be ruled by a "master race," so far in advance of other species as to have a firmness of control beyond the imagining of earthly potentates.

But even this scenario, despite its apparent neatness, would likely be unstable. As we know from our own history, the colonists sent out by any imperial society often become the sharpest competitors with that society, if not downright revolutionaries. One could easily imagine descendants of the original galactic colonizers in dispute.

But of more fundamental concern, the sheer size of the galaxy imposes an awkward constraint. Even if the "master race" has a technology so advanced that it can travel at nearly the speed of light, it still takes 100,000 years to cross the galaxy. From the standpoint of biological evolution, this is already an interestingly long time. Any extensive colonization project would take millions of years; and by that time the colonizers would no longer resemble the inhabitants of the planet they left behind. By the time envoys from the center of empire reached the distant provinces, they would have evolved into another species.

The obvious solution to this vexing problem would be to stop evolution, or at least to control it. In this regard machine intelligence offers clear advantages: it doesn't slowly evolve at the whim of genetic mutation, and it doesn't complain about rocket trips lasting thousands of years or longer. Furthermore, it is easily adapted to the rigorous environments of space, and not constrained to the protective, but infrequent, habitats of earth-like planets (or any planets, for that matter).

It has been occasionally suggested that the last billion or so years of biological history on earth have only been a prelude to the development of thinking machines, our evolutionary successors, and the ultimate repository of most of the galaxy's intelligence. We have seen that the maintenance of order in a territory as vast as the galaxy would be simpler for machines than for organic life. Thus, while we would not be surprised if our first SETI contact was with a machine, we should also not be surprised to discover that the machine is merely in the service of other machines.

In this view, it may not be particularly puzzling that we have not yet made contact. The motivation and technologies of the machines may be enormously different from our own. Their interest in, and efforts to, attract our attention may be no greater than the trouble we take to communicate with the simple plants which dwell at the bottom of the ocean.

NOTES

1. N. Horowitz, *Scientific American*, Nov. 1977, 52.
2. H. Aumann, et al. *Astrophysical Journal* 278 (1984): L23.
3. R. Terrile and B. Smith, reported in *New Scientist*, Oct. 25, 1984, 9.
4. D. McCarthy, et al., reported in *New Scientist*, Jan. 24, 1985, 36, n.
5. A. Knoll, in *The Search for Extraterrestrial Life: Recent Developments, IAU Symposium 112* edited by M. Papagiannis (Dordrecht: D. Reidel, 1985), 201.
6. Ibid.
7. F. Drake, et al. editors, SETI Science Working Group Report. Washington: NASA, 1983.
8. B. Oliver and J. Billingham, editors, *Project Cyclops: A Design Study of a System for Detecting Extraterrestrial Intelligent Life.* California: Ames Research Center, NASA, 1972.
9. T. Owen, in *Strategies for the Search for Life in the Universe*, edited by N. Papagiannis (Dordrecht: D. Reidel, 1980), 177
10. G. O'Neill. *The High Frontier: Human Colonies in Space.* New York: William Morrow, 1977.
11. M. Hart, in *Strategies for the Search for Life in the Universe*, edited by M. Papagiannis (Dordrecht: D. Reidel, 1980), 19.
12. F. Drake, in *Strategies for the Search for Life in the Universe*, edited by M. Papagiannis (Dordrecht: D. Reidel, 1980), 27.
13. S. von Hoerner, *Naturwissenschaften*, 65 (1978): 553.
14. R. Bracewell, *The Galactic Club: Intelligent Life in Outer Space*, San Francisco: W. H. Freeman, 1974.

REFERENCES

Bracewell, R. *The Galactic Club: Intelligent Life in Outer Space.* San Francisco: W. H. Freeman, 1974.
Cameron, A. editor, *Interstellar Communication.* New York: W. A. Benjamin, 1963.
Drake, F. et al. editors. *SETI Science Working Group Report.* Washington: NASA, 1983.

Figure 19-1.—Are we alone?

Goldsmith, D. editor, *The Quest for Extraterrestrial Life: A Book of Readings*. Mill Valley: University Science Books, 1980.

Morrison, P. et al. editors. *The Search for Extraterrestrial Intelligence SETI*. Washington: NASA, 1977.

Oliver, B. and Billingham, J. editors. *Project Cyclops: A Design Study of a System for Detecting Extraterrestrial Intelligent Life*. California: Ames Research Center, NASA, 1972.

O'Neill, G. *The High Frontier: Human Colonies in Space*. New York: William Morrow, 1977.

Papagiannis, M. editor. *Strategies for the Search for Life in the Universe*. Dordrecht: D. Reidel, 1980.

Shklovskii, I. and C. Sagan. *Intelligent Life in the Universe*. San Francisco: Holden-Day, 1966.

20

SETI and Its Fringe Benefits

SEBASTIAN VON HOERNER

G. Seth Shostak has presented a good summary about SETI, the search for extraterrestrial intelligence. To our discussion I would like to add first two short remarks about statistics and intelligence. My main topic then will be some spin-offs or fringe benefits of SETI, which, in my opinion, may at present be even more important than the main purpose of our search.

It is a common objection against SETI that in all the universe we know but one example for the evolution of life and intelligence, our own on earth, and one cannot draw general conclusions from a single case. The question is: *Can* we do statistics with $n = 1$? And the answer is: *Yes*, if we know the rules. A single case, $n = 1$, does give an estimator for the average, but none for the mean error (it requires $n = 2$). In plain language: the assumption that we are average has the largest probability of being right; but we have no idea how wrong this may be. To reduce this ignorance we must leave statistics and use analogy: arguing that all things in the universe do have only a finite scatter (or standard deviation) that mostly is not extremely large. Thus, the assumption that planets similar to earth gave rise to a similar evolution is quite reasonable, but not compelling.

Also, nature can reach similar results in very different ways. For example, consider flying. It has been invented independently and with five different means by: seeds, insects, bats, birds, and humans. We must keep this in mind when we multiply probabilities of intermediate steps in order to derive the probability of some result.

Regarding intelligence: is this really an asset, the crown of evolution, as we always tacitly assume, or is it just some freak of nature?

Or even a dead-end road? This question is nicely discussed by Isaac Asimov, who points out that it is the large, intelligent animals that are easily endangered by extinction, needing our protection; whereas the little dumb ones just lay a million eggs and keep going, in spite of all our efforts to extinguish some of them. It seems that fecundity has a lot more survival value than intelligence. Asimov suggests that there is a critical level, above which intelligence gives enough power to gain a spectacular advantage of dominance and safety. I agree that this seems a good answer to the question. But I would like to add another, higher critical level, above which intelligence enables self-extinction of the species—being very dangerous again.

Now to the main topic. If SETI ever succeeds, this will have tremendous impact (we merge into an old galactic culture). After a long, dedicated search, even no success is a highly dramatic answer (we are alone). But, one way or the other, SETI will most probably take a *very* long time. And meanwhile I hope we will profit by its fringe benefits: meditating about life in space may give us the proper distance and better perspective to look back to earth and our human troubles and doings.

Regarding the serious problems we have on earth, it could well be the most important thing to turn people's minds towards the stars and to the question of life in space. We are so deep in thoughts and traditions thousands of years old; most of our behavior even got bred during hundreds of thousands of years, and was developed for circumstances of the past: small groups with lots of space and unlimited resources, primitive weapons, little organization, and power. All this has drastically changed, by dangerous amounts, but we still keep going in the same old ruts. We must get out of these ruts, must start to think in different and new terms; we must find new ways in better directions. Maybe the best we can do is to stand still for a while, lift our view up from the rutted ground, and look up to the stars for better orientation, thinking about life in general, and in other places. And then, with our mind still "out there," we may turn our view back to earth with a fresh perspective, to see what must be done.

It would be healthy to make our political and military leaders meditate on this question: what will happen if the Little Green Men land on earth; they surely will be curious what to find here. Let our leaders and advisers think about it, how to explain our present state on earth to our visitors from space!

If *we* had gone to some other planet, coming down in *our* flying saucers, our most important question would be whether we find intelligent life on that planet. And this again is the Little Green Men's first question: *"Is there intelligent life on earth?"* Actually, it was SETI pioneer Frank Drake who asked this question first. We pride ourselves to be rational, intelligent beings, but a lot of strong evidence speaks against it.

The worst example is our arms race. How can people intelligent enough to design nuclear bombs be stupid enough to build them? Let me make the general statement: the largest organized human efforts have *always* been self-destructive. The only exception I can think of is the Egyptian pyramids; any other really large project, being a large part of the gross national product, was military, which is no sign of intelligence. And our arms race has acquired a life of its own, beyond rational control, senseless even in the military sense. The March 1983 issue of *Physics Today* had several articles about the arms race, with good numerical estimates from experts. The United States had a destructive power of 4,000 megatons while Russia had 6,000. On the other hand, it has been estimated how much destructive power is needed for the second strike capability, or Mutual Assured Destruction (MAD), which means that each country, after having been fully attacked by the enemy, still can strike back and successfully destroy the enemy's population centers. The answer was 400 megatons each. This means that in 1983 the U.S. had ten times the needed power, the USSR fifteen times. Yet we were tremendously worried that they had more than we did. Now both sides are still increasing and improving their arsenals, all our greatest efforts go to the military (and hopefully down the drain, never to be used). By the way, the total destructive power, 10,000 megatons, is over two *tons* of dynamite for every living person on earth—babies, grandmothers, black, yellow, white. And, for comparison: all of the shells, missiles, and bombs used in World War II (including Hiroshima and Nagasaki), had a total power of only 3 megatons.

The arms race is not the only evidence against intelligent life on earth. Just look at our standard of living: one-fourth of the human population is dying from overfeeding, one-half lives in a rather poor state, and one-fourth is dying from starvation. And look at it in connection with unemployment and population explosion (70 million more people each year).

Furthermore, look at our attitude toward the value of the human life. There is a huge fight against abortion because human life is sacred; one leading religion even forbids contraception. But most of these dedicated protectors of human life have not much against sending their soldiers to war; they do not fight against the training of grown-up men to kill other grown-ups and get killed. In my opinion, only pacifists and conscientious objectors, only those, have the right to fight abortion.

Another hopefully fruitful fringe benefit of SETI is to consider in parallel the two questions: what might be the activities of higher advanced civilizations? What should we do in our future? These two questions, if asked and answered in parallel, may on the one hand lead us to good search modes for SETI, suggesting probable types of evidence to look for. On the other hand, they may lead us to make proper plans for our own next century. Hopefully such plans could resemble those described here by Brian O'Leary in his exciting paper about the exploration and use of our Solar System. This would be a promising and productive large effort, using exactly the same technology and experts as our present insane arms race, but a lot less money.

REFERENCES

Isaac Asimov. *Extraterrestrial Civilizations.* New York: Crown Publishers, 1979. The value of intelligence is questioned, starting page 186.

S. Fred Singer, ed. *Is There an Optimum Level of Population?* New York: McGraw-Hill, 1971. Extensive discussion of population explosion, food, natural resources, and quality of life.

Physics Today 36, no. 3, March 1983. Useful articles about nuclear deterrence and the existing nuclear arsenals.

PART IX

UFOs: FANTASY OR PRESENT REALITY?

21

UFOs: Fantasy or Present Reality?

BRUCE MACCABEE

Introduction

> This "flying saucer" situation is not all imaginary or seeing too much in some natural phenomenon. Something is really flying around.[1]

Reports of unusual objects moving around in the sky have been made occasionally since biblical times.[2] Therefore it is not particularly surprising that unusual objects are still being reported in the skies, although there does appear to be a major difference between the ancient and the modern reports. Generally the ancient reports could rather easily be explained as misinterpretations of natural phenomena that have only recently become understood or as "visions" (psychological manifestations, often of a religious nature). However, a considerable number of the modern reports that have been made since the spring of 1947 are different from the ancient reports because they seem to be inexplicable, *even after careful investigation and analysis,* in spite of our increased knowledge of natural phenomena and our increased technological sophistication. The reports I refer to form a subgroup of the general class known as "flying saucer" or UFO (Unidentified Flying Object) reports.[2] I call the members of this subgroup TRUFO (TRue UFO) reports. Examples are presented in this paper.

The "Generalized UFO Phenomenon" can be defined as the continuous flow of reports, from throughout the world, of sightings of aerial objects or phenomena that the witnesses cannot identify. An important characteristic of the generalized UFO phenomenon is its persistence or "robustness" in the face of the skeptical attitude toward such sightings on the part of the general population and scientists in particular. Reports continue to be made by persons who would, under other circumstances, be considered reliable and reasonably accurate observers. The reports continue in spite of a generally skeptical attitude that often leads to ridicule if a report becomes public knowledge.[3] (Note: Although reports by mystically or religiously oriented persons, by hoaxers and by charlatans are also part of the "Generalized UFO Phenomenon," as a physical scientist I don't consider them to be an important part, although a sociologist might.)

It is natural to suspect that such a robust phenomenon has a real, hitherto unknown underlying cause, whether it is purely psychological in origin or is based on physical reality. In other words, it is natural to suspect that at least *some* reports result from sightings of phenomena that remain unexplained after careful investigation, just as it is natural to suspect that "where there's smoke, there's fire." I label as "TRUFOs" those phenomena that remain unexplained after careful investigation of reports and after subsequent expert analyses of the information obtained from the reports. It is important to understand that the investigation of a sighting concentrates on establishing the level of credibility of the *description* of the phenomenon as given by the witness(es) rather than on the witness' own interpretation. The analysis of a sighting is also based on the witness' description rather than on the interpretation.

In order to find out whether or not any TRUFOs have been sighted, it is necessary to study very carefully many individual sightings. Such a study will show that *most* sightings can be explained in terms of known, if rare, phenomena or psychological manifestations (including hoaxes). However, some sightings will be found to defy logical explanation in terms of known phenomena. The question, then, is this: why can't each these sightings be explained? Is it because each of the unexplained sightings was made by a poor observer and the report doesn't provide enough high quality information for a positive (or even probable) identification? Or is it because the quality, self-consistency, and apparent accuracy of each report make it very difficult or impossible to conceive of any rational explanation in terms of known phenomena?

TRUFOs or Insufficient Information?

"Of course there always will be UFOs."

A statement like this is sometimes made by a skeptic who realizes that there will always be some sightings that can't be explained because of a lack of information. For example, a UFO investigator might fail to identify a reported object as a balloon because he did not know that a balloon could have been at the location of the sighted object. As another example, a sighting of Venus might go unrecognized because the witness failed to note the time or direction of the UFO.

Is it true that the unexplained sightings are simply those lacking sufficient information for a definite explanation? The answer to this question is no, according to an Air Force sponsored[4] study that was done many years ago by the Battelle Memorial Institute of Columbus, Ohio. Their scientists worked on Project Stork for the Air Force during the years 1952 and 1953.[3, 5, 6] Project Stork was a classified (secret) in-depth study that introduced state-of-the-art (at that time) computer-aided statistical analysis techniques into the field of UFO research ("UFOlogy"). The BMI scientists, along with the Air Force experts, studied 3,201 reports that had been collected by the Air Force from 1947 through 1952. By carefully analyzing the descriptions of the reported phenomena they found that 2,214 reports (69.2 percent) could be explained with a reasonable degree of confidence. For purposes of statistical analysis the explained reports were designated "K" for "Known."[7] Reports that didn't provide sufficient information to allow for a reasonably confident identification of the phenomenon were labeled I.I., for "Insufficient Information" (298 reports or 9.3 percent). The scientists probably were not[4] surprised to find a sizable number of reports (689 or 21.5 percent), which they couldn't explain even though the descriptions appeared to be quite clear, complete, and self-consistent. They labelled these "U" for "Unknown". *They emphasized that the U reports were clearly distinct from the I.I. reports.*[5]

TRUFOs or Poor Observers?

> Less well informed individuals are more likely to see an UFO. . . .
> In our experience those who report UFOs are often very articulate but not necessarily reliable.
>
> —E. U. Condon[2]

The BMI scientists analyzed each sighting to determine an overall estimate of its reliability. They constructed a semi-quantitative scale of the reliability of a sighting by combining a relative scale value of the reliability(ies) of the witness(es) with another relative scale value of the reliability of the report. The relative reliability of a witness was based on his (her) experience as "deduced from his occupation, age, and training" and upon the "observer's fact reporting ability and attitudes, as disclosed by his manner of describing the sighting."[5] The relative reliability of a report was based on "consistency among the several portions of the description of the sighting" and upon the "general quality and completeness of the report."[5] Once the overall relative reliability had been determined for each report the scientists divided the reports into four levels of relative reliability, Poor (525 or 16.4 percent), Doubtful (1298 or 40.5 percent), Good

(1070 or 33.4 percent) and Excellent (308 or 9.6 percent).[8] They then calculated the percentages of K, I.I., and U sightings for each reliability level. They found that the percentage of I.I. sightings generally decreased as the reliability increased from Poor to Excellent (19.6 percent, 11.6 percent, 3.1 percent, and 3.9 percent, respectively). This was expected since the relative reliability assigned to a sighting depended upon the amount of credible, self-consistent information contained within the report. They also found that most sightings within each reliability level could be explained (62.1 percent, 72.8 percent, 70.6 percent, and 61 percent, respectively). What wasn't expected was that the percentage of unexplained sightings generally *increased* with reliability (18.3 percent, 15.6 percent, 26.3 percent, and 35.1 percent, respectively). The percentage of unexplained Excellent sightings increased even further when civilian sightings were removed from the total collection leaving only military sightings, most of which had been made by servicemen on duty at the times of the sightings. Of the 1,226 military sightings, 204 (16.6 percent) were rated as Excellent and of these 37.7 percent were sightings of Unknowns according to the Battelle scientists.[9]

I have pointed out elsewhere that, if there were no TRUFOs to be sighted (i.e., TRUFOs don't exist), then this result implies that the more reliable a witness is, the more likely he is to make mistakes in his report of a sighting.[6] On the other hand, if there were TRUFO sightings in the BMI sample, then the result is completely consistent with our understanding of reliability.[10]

The Air Force did not mention this conclusion about the relationship between reliability and the percentage of unexplained sightings when the report was declassified and released in 1955.[3] Nor did the Air Force publicize Battelle's conclusion that 21.5 percent of all of the sightings made between 1947 and 1952 had not been explained.[3]

I have studied many UFO reports and have found a number of them that appear to me to have more than enough credible information for identification if identification were possible. These are, in my opinion, reports of TRUFOs (TRUFO reports). Some examples are discussed in later sections of this paper.

I therefore tend to agree with the opinion presented at the very beginning of this paper, an opinion that was written by an Air Force intelligence analyst in the summer of 1947: "Something is really flying around."[1]

Early Air Force Opinion

"It can't be, therefore it isn't."

This is the way Dr. J. Allen Hynek, former consultant in astronomy to the Air Force during Projects SIGN and BLUE BOOK, has characterized the attitude taken by the Air Force investigators, especially in later years.[11] However, it wasn't always that way. Initially Air Force intelligence concluded that at least some reports were of real objects. The opinion stated at the beginning of this paper[1] was repeated by Lt. General Nathan Twining in a Sept. 1947 letter to Brig. Gen. George Schulgen, who was the Chief of the Intelligence Requirements Branch of (Army) Air Force Intelligence.[12] Twining indicated to Schulgen that there was a requirement for a special project to analyze flying saucer reports. The Air Force responded to Twining's letter by setting up Project SIGN.[13]

The opinion that real objects were the causes of some UFO sightings was also expressed in a formerly Top Secret Air Intelligence document that was written in December 1948 but not declassified and made available to the public until *March 1985*. This document expressed the conclusion that "some type of flying objects have been observed, although their identification and origin are not discernible."[14]

This recently released document appears to be a rewritten or "watered-down" version of the legendary "Estimate of the Situation," which was written by the staff of the Air Technical Intelligence Center (now the Foreign Technology Division) at Wright-Patterson Air Force Base in the late summer of 1948.[15] According to Capt. Edward J. Ruppelt, the first director of Project BLUE BOOK, the Estimate presented the conclusion that flying saucers were extraterrestrial vehicles. The estimate is "legendary," since all official copies were destroyed in the years following General Hoyt Vandenburg's rejection of its conclusion.[3, 15] (Vandenburg was the Chief of Staff of the Air Force.)

In view of the skeptical attitude toward UFO sightings that was expressed publicly by scientists and also by the Air Force in 1947 and 1948, one wonders: what could have made the intelligence analysts confident that "some type of flying objects" had really been observed? In view of the skeptical attitude of the Air Force and of the general scientific community through the 50s, 60s, and 70s and thus far into the 80s, the reader may well wonder why *I* tend to agree with the opinion of the Air Force Intelligence analysts in 1947–1948. After all, hasn't this question of UFOs been "put to sleep" by numerous scientists from various fields over the past thirty-five years? Consider, for example, the Colorado University Study[2], or the papers by Urner Liddell[16], William Markowitz[17], and Donald Warren[18], or the books by Donald Menzel[19, 20, 21], Philip Klass[22, 23, 24], Robert Sheaffer[25], and James Oberg.[26] Haven't these works established that "further extensive study probably cannot be justified in the expectation that science will be advanced thereby?"[2] Haven't they shown that there are no truly unexplainable reports?

These questions can best be answered after a study of some representative TRUFO reports. These reports provide a basis for understanding the claim that "something is really flying around."

TRUFOs by the Dozen

"I have a little list. . . ."[27]

In Table 21-1 I have listed about two dozen reports that I believe are a "challenge to science."[28] For each one, there is enough information available to allow for an explanation if an explanation were possible, yet there seems to be no reasonable explanation. I have chosen these out of the thousands available for a combination of reasons. One reason is the interesting nature of the reported phenomenon. A major reason for choosing some of them has been my own involvement in the investigations. I have directly investigated, either through "paper research" or through personal interviews with witnesses, several of the sightings (numbered 6, 7, 8, 9, 13, 19, 20, 21, 22, 23, 24, 25). Another reason is that a number of the sightings have been known for many years and have been critically discussed by several authors. Some have even been "explained," often in several different ways. I have learned how avid skeptics have treated the UFO problem by studying these "explanations." In particular, I have found that in many cases the explanations appear to be wrong or at the very least unconvincing. I have also selected these cases to cover modern UFO history and to illustrate the types of sightings that have been reported. Finally, I have chosen these cases for their high levels of physical detail and witness credibility, as compared to reports of distant lights seen at night and distant objects seen during the daytime without optical aids or to reports of abductions and interactions with "UFOnauts." I have chosen these reports because the witnesses appear not to have had any "axes to grind" in making the reports. Instead, they are ordinary people who have been involved in extraordinary events for short periods in their lives and then have reverted to ordinary behavior after the events. I have not included reports made by people who claim repeated contacts with or ascribe religious purposes to "UFOnauts" (e.g., Adamski, Bethuram, Fry, etc.; see note 3).

Needless to say, this particular collection is not unique. Someone else who selects a limited number of sightings for detailed analysis might choose an entirely different set. For example, Dr. Thornton Page selected sixteen cases for illustration in the book *UFOs, A Scientific Debate*, which reported on the AAAS UFO meeting that was held in December 1969.[29] There is no overlap between his selection and the list in Table 21-1. On the other hand, Ronald Story's selection of ten cases[30] does include several that are presented here (numbers 14, 16, 22, and 23).

The reports in Table 21-1 have been discussed at greater length in an unpublished paper that is over a hundred pages long.[31] Because of a lack of space, only about a dozen are discussed in this paper. Nevertheless, these few examples should suffice to illustrate that

TABLE 21-1.
TRUFO REPORTS

Date	Type of Sighting	Witness(es)
1) Apr. 1947	visual; theodolite	meteorologists
2) June 24, '47	visual from aircraft	private pilot
	visual from ground	prospector
3) July 8, '47	visual	MUROC Air Force Base employees and officers
4) July 8, '47	material recovered	farmers, A.F. Major
5) Apr. 24, '49	visual; theodolite	meteorologists
6) May 24, '49	visual	Ames Research Lab. engineers
7) Jan. 22, '50	visual; radar	Navy pilots
8) Apr., May '50	cinetheodolite films	Army missile scientists
9) May 11, '50	visual; photograph	farmers
10) Aug. 15, '50	visual; movie	manager/secretary
11) Oct. 10 & 11 1953	visual	Skyhood baloon scientists
12) May 5, '53	visual; optical effect	chemist
13) Aug. 29, '56	visual; photographic	Canadian AF pilot
14) Nov. 2–3 '57	visual; effects on automobiles	multiple
15) Jan. 16, '58	visual; photographs	multiple (Trindad Is.)
16) June 26–28 1959	visual	multiple (Gill/Boainai Is.)
17) Apr. 24, '64	visual	police officer (Zamorra)
18) Oct. 21, '65	visual; photo	deputy sheriff (Strauch)
19) Aug. '66	visual	Navy chemist, Navy pilot
20) Apr. 17, '70	visual	elected official
21) July 27, '75	visual	Navy technical writer
22) Sept. 21, '76	visual; radar	Iranian AF jet pilots
23) Dec. 31, '78	visual; radar; movie film; tape recording	multiple witnesses (New Zealand)
24) Aug. 9, '80	visual	security guards (Kirtland AFB)
25) Dec. 4, '81	visual; photographic	multiple

there are sightings with credible information reported by qualified observers, and that skeptics have sometimes incorrectly applied methods of scientific analysis in order to provide explanations for some sightings. One may even infer, from the military sightings, that the U.S. Air Force still has an involvement with the UFO problem.

Discussion of a Report and the Explanations

They were flat and flew like "pie plates skipping over the water."
—Kenneth Arnold[2]

Kenneth Arnold's sighting on June 24, 1947, was the first publicized sighting, but it was not the first sighting recorded by the Air Force for the year 1947. The date of the first sighting was January 16; the location was over the North Sea. On that night, at 10:30 P.M., a British Mosquito aircraft chased an unidentified object for half an hour. According to the official report, "the unidentified aircraft appeared to take efficient, controlled evasive action." The British government did not report this incident to the U.S. (Army) Air Force until July because the Air Force had indicated no interest in such sightings until after Arnold's sighting.[32]

Several sightings were made during the first three months of 1947 by meteorologists in Richmond, Virginia. On four occasions, each time while tracking a weather balloon, a meteorologist saw a "strange metallic disc" fly past the balloon. The observations were made through a theodolite telescope.[14] During one sighting, in April, a "disc" was tracked for fifteen seconds as it traveled from east to west on an apparently level track. To quote from the report, "The object was a metallic-like chrome, shaped something like an ellipse with a flat level bottom and a dome-like round top. The disk appeared to be below the balloon, was much larger [than the balloon] in [angular] size in the [field of view of the] theodolite and shined like silver." The balloon was at fifteen hundred feet at the time. The visibility was good, with generally clear skies. This report was chosen by the writers of the Air Intelligence analysis[14] as one of the most important examples of good quality sightings from experienced observers.

These sightings, as with the British sighting, were not reported to the Air Force until July. The reason that the meteorologists did not report their sighting immediately was that they thought they were seeing a new type of U.S. aircraft. However, in July the Air Force stated publicly, with support by statements by famous scientists such as Vannevar Bush, Merle Tuve, and David Lilienthal[3, 15], that there were no U.S. experimental aircraft that could account for "flying disk" reports. When the meteorologists learned this from press reports they concluded "we must assume this strange object to be foreign" and submitted their report to the Air Force "for yourjuinformation."

There were several other notable but unreported sightings before June 24, 1947. However, Arnold's sighting alerted the nation to the fact that people were seeing strange objects flying around in the skies, and within days local newspapers and national news services were reporting sightings made both before and after Arnold's sighting.

A careful study of Arnold's sighting is valuable for several rea sons: it is the prototype of daylight sightings by aircraft pilots; it was the first publicized report and therefore could not have been made as a response to any publicity; it was made by a well-respected and credible observer (Arnold was a businessman with over four thousand hours of flying time); Arnold was interviewed by Air Force investigators and found to be credible; Arnold sent to the Air Force a complete description of his sighting and how it came about; and finally, there are numerous published "explanations" that illustrate how skeptics deal with reports from credible observers. This latter reason is of particular interest to me since I have been interested in studying the techniques used by skeptics to "explain" UFO reports.

Arnold's sighting report is summarized in a number of books.[2, 3, 11, 14, 15, 19, 20, 21] Therefore the complete story will not be presented here. However, certain statements, extracted from Arnold's written report to the Air Force[32], will be mentioned here as they are needed in order to evaluate the explanations that were offered.

Numerous scientists, science writers, and laymen offered explanations for the sighting but the people who had the most impact on the final Air Force evaluation were Dr. J. Allen Hynek and Dr. Donald Menzel. Moreover, their explanations include all of the potentially valid suggestions offered by others. Therefore I will discuss only their explanations.

In 1947 the Air Force intelligence analysts decided that Arnold's sighting as unidentified, and it is listed as such in the Air Intelligence study[14] mentioned previously. About a year later Hynek analyzed the sighting as part of his work for Project SIGN[32, 33] and pointed out an inconsistency in the report.

Arnold had claimed that while he was flying eastward from Mineral, Washington, in the middle of the afternoon of June 24, 1947, he saw nine thin, semicircular (crescent shaped) objects that flew along a south-southeasterly course past Mt. Rainier. They tilted back and forth as they flew, causing the sun to repeatedly flash in his direction. The objects "swerved in and out of the high mountain peaks" south of Mt. Rainier. They traveled one after the other in a line that Arnold estimated, by comparison with known mountain peaks, to be about five miles long. They were twenty to twenty-five miles away at the time he saw them from his airplane and he thought that they were about the size of a jet fighter (45 to 50 ft). Using the dashboard clock in his airplane Arnold measured the time from when the first object passed Mt. Rainier until the last object passed Mt. Adams, a distance of about forty-seven miles. It took 102 seconds. Using this information Arnold estimated their speed at over 1,500 mph.

Hynek began his analysis of the sighting by calculating the size of the objects. Arnold had said that he could see them edge-on and that he thought their width was about twenty times greater than their thickness. Assuming three minutes of arc as the minimum angular size to be visible edge-on, Hynek concluded that if the objects had been twenty-five miles away they would have been about 100 ft thick and 2,000 ft long. Hynek then pointed out that if they had been only 45 to 50 ft in size, as estimated by Arnold, they would have been much closer (Hynek estimated six miles away) in which case their calculated speed could have been lower (400 mph). Based on this reasoning Hynek suggested that Arnold's report was self-inconsistent and that he had overestimated the distance. Hynek suggested that Arnold had seen ordinary aircraft.

As a result of Hynek's discussion of the discrepancy between Arnold's estimates of the distance and size of the objects, the Air Force decided that "the entire report of this incident is replete with inconsistencies" and cannot bear even superficial examination" (from the Project Grudge Report).[32]

Evidently Hynek had failed to pay careful attention to Arnold's report, which states that the objects traveled very close to Mt. Rainier and close to the peaks of the mountains that form a chain lying between Mt. Rainier and Mt. Adams. Arnold stated in his report that he was quite sure of the distance of the objects because as they flew along the mountain chain "there were several peaks that were a little this side of them as well as higher peaks on the other side of their pathway." At the time of the sighting Arnold had been flying for two or three minutes eastward toward Yakima, Washington, from Mineral, Washington, which is about twenty-four miles west-southwest of Mt. Rainier. This unequivocally establishes the minimum distance of the object at about twenty miles. A recent recalculation based on this distance, based on the assumption that Arnold overestimated the width to thickness ratio by more than a factor of two, and based upon a better estimate of the capabilities of the eye, suggests that the width of the objects was probably much less than Hynek's estimate of 2,000 ft.[31] The speed was actually about 1,800 mph.

Hynek's work was done secretly for the Air Force. However, a few civilian scientists had access to Air Force files and one of these, Donald Menzel, decided to write about Arnold's sighting in his first book on UFOs[19]. He criticized the Air Force for accepting Hynek's explanation. He stated: "Although what Arnold saw has remained a mystery until this day, I simply cannot understand why the simplest and most obvious explanation of all has been overlooked . . . the association of the saucers with the hogback (of the mountain range) . . . serves to fix their distance and approximate size and roughly confirms Arnold's estimate of the speed." Menzel then went on to suggest that Arnold saw "billowing blasts of snow, ballooning up from the tops of the ridges" caused by highly turbulent air along the mountain range. According to Menzel, "These rapidly shifting, tilting clouds of snow would reflect the sun like a mirror . . and the rocking surfaces would make the chain sweep along something like a wave, with only a momentary reflection from crest to crest."

This first explanation by Menzel may seem slightly convincing, but only until one realizes (a) that the sighting took place at 3:00 P.M.

when the sun was high in the sky west of Arnold, (b) that snow cannot reflect light rays from the overhead sun into a horizontal direction "like a mirror," (c) that there are no 1,800 mile winds on the surface of the earth to transport clouds of snow or to create an oscillating wave with a phase velocity of 1,800 mph, and (d) that there are no winds that would carry clouds of snow all the way from Mt. Rainier to Mt. Adams (Arnold saw the objects fly past Mt. Adams before they faded from view.) Furthermore, if Menzel were correct, one wonders how Arnold could have failed to realize that he was seeing snow blowing from the mountain tops, especially since he actually flew past Mt. Rainier and therefore would have gotten much closer to the mountain within minutes after his sighting.

Menzel also offered a second possible explanation for the sighting: there was a thin layer of fog, haze, or dust just above or just below Arnold's altitude, which was caused to move violently by air circulation, and which reflected the sunlight.[19] Menzel claimed that such layers can "reflect the sun in almost mirror fashion." Menzel offered no substantiation for this claim. Perhaps he was thinking in terms of a "reflection" from an atmospheric layer when the sun is at a grazing angle (near the horizon). If so, then that explanation makes no sense since the sun was nearly directly overhead (and slightly behind Arnold) at the time of the sighting. Again, one wonders how Arnold could have failed to notice that he was just seeing the effects of a haze layer.

Menzel explained Arnold's sighting a third way in his second book[20]. He suggested that the objects were actually mirages of the mountain tops. To support this explanation he presented a photograph of such mirages. Such mirages can be seen when the observer's line of sight to the mirage is tilted no more than one half of a degree above or below the observer's horizon. Evidently Menzel overlooked the following information in Arnold's report: Arnold reported that the objects were about at his height (he estimated their altitude at 9,500 ft; his altitude was 9,200 ft), but that he saw them *silhouetted against the side of Mt. Rainier, which is 14,400 ft high.* Thus the objects were far below the location of any mirage of the top. Moreover, the angular elevation of the peak of Mt. Rainier from Arnold's location would have been over two degrees, far too great for a mirage. Of course, such mirages stay above the mountain peaks, so the mirage theory can't account for the high lateral speed reported by Arnold.

Menzel also suggested a fourth explanation: wave clouds in rapid motion.[20] This explanation was supported by a "sighting" and a photograph of such a cloud by a newspaper photographer. This explanation also fails to account for the bright reflections reported by Arnold, for the distinct shapes, or for the high speed that Arnold measured.

In his third and last UFO book, which is subtitled *The Definitive Explanation of the UFO Phenomenon,* Menzel again discussed Arnold's sighting and offered yet a fifth explanation: Arnold saw water drops on the window of his aircraft.[21] To support this claim Menzel offered an account of a "sighting" of his own that turned out to be water drops condensed on the outside of the aircraft window. The water drops were so close to his eye that they were out of focus and he thought they were distant objects. They moved slowly along the window surface from front to back, given the illusion of distant objects moving at great speed . . . until he refocused his eyes and discovered what they were. In comparing his "sighting" with Arnold's Menzel states: "I cannot, of course, say definitely that what Arnold saw were merely raindrops on the window of his plane. He would doubtless insist that there was no rain at the altitude at which he was flying. But many queer things happen at different levels of the earth's atmosphere." Had Menzel bothered to reread Arnold's report before writing this "explanation" he, presumably, would not have proposed the water-drop explanation because he would have found Arnold's statement that he turned his plane sideways and viewed the objects through an open window.

Incidentally, the Air Force eventually decided that what Arnold saw was a mirage.[32] My own opinion is that he saw TRUFOs.

An Excessive Urge to Explain

Skeptics sometimes claim that an inner desire for "the unexplained" keeps UFO proponents from accepting reasonable explanations for sightings. No doubt this is at least partially true for some proponents. However, it appears to me that the reverse is also true: it seems that some skeptics have such a great need for explanations that they will accept faulty explanations, and explanations that conflict with the sighting report. The Arnold sighting is a case in point.

In reviewing Hynek's and Menzel's treatment of the "first" UFO sighting I was struck by three factors: (a) both men accepted Arnold's sincerity and considered him to be a credible, reasonably accurate witness; neither man thought Arnold's sighting was a hoax or a manifestation of some psychological problem; (b) both men made use of only certain parts of the information Arnold supplied, those parts which tended to support their hypotheses; and (c) when one explanation seemed unconvincing, another was suggested. Specifically, Hynek accepted Arnold's estimate of the size of the objects, yet rejected the estimate of distance, even though the distance estimate was more likely to be correct since the objects flew near known landmarks (mountain peaks). Menzel, on the other hand, accepted the distance and the high speed. He then conceived of explanations in terms of natural phenomena, and either extrapolated the characteristics of those phenomena far beyond the known characteristics, or else he ignored certain details in Arnold's report when they didn't fit the explanation. For example, Menzel extrapolated the light-bending capability of a mirage far beyond what is known to occur. He ignored the fact that mountain top mirages remain over the mountain tops. He attributed specular reflectivity (mirror-like reflection capabilities) to clouds and to dust layers in the atmosphere.

From comparing the explanations of this sighting with explanations of other sightings I have studied, I have concluded that certain people have an excessive urge to explain. They propose explanations that do not stand up to scrutiny. Yet, they are entirely serious in claiming that the sightings have been explained and, when one explanation is called into doubt, they offer another. Menzel, in this case, offered five explanations in his three books, when one should have been sufficient.

It is as if some skeptics feel that the probability that a sighting can be explained increases with the number of explanations that can be offered. Actually the probability is either independent of the number, or perhaps it even decreases as the number of explanations increases. I have often felt that whenever several different experts offer several different explanations for a sighting, then it probably has not been explained after all.

A number of further sightings are described in the notes available from the author. The descriptions are based on information from several sources and include discussions of explanations that have been proposed. They provide further illustrations of the type of well-reported sightings that exist and of the attempts that have been made to explain such sightings. The reader will note that two skeptics in particular are mentioned several times (Menzel and Klass). These men occur most often because, unlike most skeptics, they have published their explanations in books that are likely to be read by scientists who are more or less casually looking into the subject. Furthermore, they have presented explicit, sometimes semi-quantitative explanations that can be compared with the information available in the sighting reports, thereby allowing for a reasonably definite judgment as to whether or not their explanations are acceptable.

TRUFOs, "The Modern Myth"

Skeptical scientists have long argued that TRUFOs are not real; that they are a "modern myth."[34] This argument is based, ultimately, on the claim that all sightings can be explained. Obviously that claim is not universally accepted (if it were, this paper wouldn't exist). Why not? Where lies the problem with universal acceptance of the idea that all sightings can be explained?

As I see it, the problem does not lie with the convincingly explained sightings. Nor does it lie with those sightings that have not been explained because of a lack of information (it is possible to decide when a report does not provide enough information about the phenomenon or about the circumstances under which the sighting took place). Instead, the problem lies with those well-reported sightings for which the published explanations have, after careful analysis, been found to be wrong or simply unconvincing. When wrong or unconvincing explanations are proposed for well-reported sightings by credible observers, less skeptical scientists and educated laymen may conclude, rightly or wrongly, that these sightings *can't* be explained.

Skeptical of the Skeptics

Since "explanation" lies at the heart of science, skeptics have a fundamentally appealing argument when they claim that all sightings can be explained. Therefore, a scientist who is just beginning a study of the UFO problem is likely to initially accept the skeptical point of view. This acceptance will lead the scientist, as it led me, to try to explain sightings.

I like explanations when they are convincing, and especially when they are correct. (Note: one cannot always know whether or not an explanation is correct, but one can know, to within a reasonable degree of confidence, whether or not an explanation is convincing.) I like to make a UFO sighting "go away" as a result of a good explanation. I always begin a sighting investigation by assuming that it can be explained. Moreover, I always immediately start forming "candidate explanatory hypotheses" as soon as the data (sighting information) become available. These hypotheses help me formulate key questions that can identify the true nature of the reported phenomenon. I have provided a number of explanations for visual and also for photographic "sightings." Thus, I accept the general idea of trying to explain UFO sightings and, to that extent, I agree with the skeptics. It was only after a careful study of some published explanations, followed by my realization that the explanation just could not be correct, that I became skeptical of the skeptics.

From Whence Cometh . . .

If TRUFOs are real, a lot of room for speculation is opened up. Although I tend to take the experimental approach and let the data "speak for themselves" (the UFO data have not spoken clearly yet), I will make some brief comments on theories that presume that at least some UFOs sightings have resulted from a phenomenon (or phenomena) that is (are) inexplicable in terms of conventionally accepted phenomena.

Some theories suggest that some TRUFOs are hitherto unknown or not well understood natural *unintelligent* phenomena. For example, some sightings of odd lights might be sightings of ball lightening.[22, 30] However, theories that attribute all unexplained sightings to natural phenomena must explain how natural unintelligent phenomena can look like "craft" or "machines" and can behave dynamically as if responding to human actions. Natural unintelligent phenomenon theories also have difficulty explaining reports of "entities" or "UFOnauts."

Other theories assume that some intelligence lies behind UFO reports. The intelligence could be either human or nonhuman. One "human intelligence" theory is that TRUFOs are a "secret development" of some government or secret organization.[30] The history of sightings since 1947 and the government reaction to them make this theory nonviable. Another such theory is that TRUFOs are psychological constructs within the witnesses. If this is so, then psychologists have a great deal of work ahead of them. This theory would have a difficult time explaining multiple witness sightings, landing trace sightings, photographic sightings, and radar sightings. Psychologists who have studied the problem have decided that there must be real unknown physical phenomena underlying at least some of the sightings.[61, 62]

Still other theories tread a fine line between physical and psychological, suggesting "projections" into our time and space or into our minds.[30] (These theories have difficulties in explaining physical effects.)

The most widely known (I hesitate to say, popular) theory is the somewhat overemphasized Extraterrestrial Hypothesis or ETH.[30, 63, 64] This hypothesis is that aliens from outside our Solar System have figured out how to get "here" from "there" in spite of astronomical distances. The ETH has inspired a number of theories that could explain how "they" got here from there without violating known laws of space and time, as well as a number of theories that violate space-time as we presently understand it. I do not intend to go into these, other than to point out two things: (a) the most credible reports in the open (i.e., unclassified) literature provide no clear evidence as to the origin(s) of the TRUFOs, and (b) arguments against brute force space travel (i.e., with rockets) are not arguments based on physical laws; instead, they are arguments based on presumed alien sociology or the desire of aliens to travel. In other words, it is incorrect to argue that they aren't here because "they can't get here from there." The fact is that if you have enough time and don't mind being cut off from your "roots," you (and your progeny) can travel immense distances. Nevertheless, the ETH has generally been ruled out by astronomers and others because of the large distances involved.[63, 64] This is a main reason why scientists have argued that UFO sightings cannot be evidence of alien "craft" and consequently, have paid little attention to the UFO phenomenon. On the other hand, it has been pointed out that this failure to take UFO reports seriously might actually lead us to overlook a real alien landing.[65]

For my part, I have tended to stay away from theories and concentrate on the "data." I feel that, ultimately, the data will stand or fall on their own merits, quite independently of theoretical arguments. I presume that once sufficient data are available it will be possible to determine "the answer" without resort to theoretical constructs (but I could be wrong on this; some physical data cannot be understood without resort to complex theoretical arguments).

Conclusion: Fantasy or Reality?

I believe that there have been for many years sufficient data (i.e., well-reported sightings) to settle the question of TRUFO reality. The challenge to science is to prove that the sightings can be convincingly explained in conventional terms. This does not mean, as some critics have suggested, that scientists must explain *all* sightings. Explanations are needed only for those sightings that clearly have scientific merit and credibility. The ones discussed in this paper would be a good place to start. Finally, if convincing explanations are not forthcoming, the scientific community should begin to seriously consider the implications of TRUFOs.

NOTES

Several of the following references can be found in the Project *Blue Book* File that is stored at the National Archives. It is comprised of records (sightings, analyses, memoranda) that were collected by Project *Blue Book* from 1947 through 1969. It also contains UFO investigations that had been carried out by the Air Force Office of Special Investigations. Microfilm copies of the complete file are available on special order. Information found in the files of other agencies has been obtained under the Freedom of Information Act regulations. The documents cited here that are not on microfilm are available from the Fund for UFO Research, Box 277, Mt. Rainier, MD 20712. An appendix to the present paper on TRUFO sightings is also available from the same address. References 35 to 60 refer to the appendix.

1. Air Force document found in the FBI "Flying Disc" file; author unknown; probable date: late July or early August 1947.

2. D. Gilmour, ed., *The Scientific Study of Unidentified Flying Objects*, chap. 5, sec. 1; AFOSR contract study F44620-67-C-0035; E. Condon, director, 1968; New York: Bantam Books Edition, 1969, 481.

3. D. Jacobs, *The UFO Controversy in America*, Bloomington, IN: Indiana University Press, 1975.

4. Memorandum in the *Blue Book* by Capt. E. Ruppelt, Director of Project *Blue Book* during 1952 and 1953. The memorandum states, in part: "It is very reasonable to believe that some type of unusual object or phenomenon is being observed as many sightings have been made by highly qualified sources" (National Archives microfilm).

5. *Special Report # 14* and the preceding Status Reports of Project Stork (National Archives microfilm).

6. B. Maccabee, "Historical Introduction to Special Report #14," Center for UFO Studies (1979) and "Scientific Investigation of Unidentified Flying Objects, Part II," *UFO Studies 3*, 24 (1983) (Both are available from the Center for UFO Studies, 1955 Johns Drive, Glenview, Illinois, 60025.)

7. The Battelle scientists had several classifications for the reports that were labeled "known": balloons, astronomical objects, optical (light) phenomena, birds, clouds, dust, psychological (including hoaxes) and "other," which included kites, flares, rockets, contrails, fireworks, etc.

8. *Special Report #14*, Table A25

9. Ibid., Table A46

10. A. Hendry, *The UFO Handbook*, Garden City, N.Y.: Doubleday and Co., 1979. This conclusion is based on the following argument: If there were no mistakes made by the witness in their reports (that is, if they gave perfect *descriptions* of the phenomena

even though their *interpretations* may have been faulty), and if the reports contained sufficient information for identification and *if there were (are) no TRUFOs*, then the sightings could be explained by experienced analysts who had access to supplementary data (balloon launchings, aircraft flight plans, weather data, etc.). The likelihood that an experienced analyst would fail to explain a sighting would increase as the number of errors in the description increased or as the amount of information about the phenomenon decreased. On a statistical basis one would expect that reports with more than average descriptive reliability, i.e., the "best reports," would contain more than the average amount of reliable information about the sighted phenomena, *and thus would be more likely to be explained*, than would reports with less than the average amount of descriptive information from witnesses with less than the average level of reliability (the "worst reports"). If, in such a statistical study, it is found that a larger fraction of the best reports remains unexplained than of the worst reports, then this can be attributed to either an increase in the number of errors within the best reports, as compared to the worst reports, or to truly unexplainable phenomena that are described in the reports and that are more likely to be recognized as such in the best reports than in the worst reports.

11. J. Allen Hynek, *The UFO Experience*, New York: Henry Regnery and Co, 1972.

12. D. Gilmour, op. cit., Appendix R, p. 894. Lt. Gen. Twining wrote that, after conferring with several Air Force technical laboratories and Air Force Intelligence, he was of the opinion that "the phenomenon reported is something real and not visionary or fictitious." Twining gave a general description of the reported objects and their flight patterns and recommended that a special project be set up to carry out a "detailed study of this matter."

13. Ibid., p. 896. As a result of Twining's recommendation, the Brig. Gen. Schulgen's agreement, Maj. Gen. L. Craigie, in Dec. 1947, ordered Project Sign to be set up at Wright-Patterson Air Force Base under Gen. Twining. Project Sign was the first of three publicly-known Air Force projects that collected UFO reports. The two follow-up projects were known as Grudge, which ran from Feb. 1949 through March 1952, and Blue Book, which ran from the end of Project Grudge through Dec. 17, 1969

14. Air Intelligence Report #100-203-79, *Analysis of Flying Object Incidents in the U.S.*, Directorate of Intelligence (of the Air Force) and Office of Naval Intelligence, 10 Dec. 1948; classified TOP SECRET until declassification on March 5, 1985.

15. E. Ruppelt, *The Report on Unidentified Flying Objects*, Garden City, NY: Doubleday and Co., 1956; New York: Ace Books, 1956

16. U. Liddell, "Phantasmagoria or Unusual Observations in the Atmosphere," *J. Optical Soc. of America* 43, no. 314 (1953).

17. W. Markowitz, "The Physics and Metaphysics of Unidentified Flying Objects," *Science*, 15 Sept. 1967, 1274.

18. D. Warren, "Status Inconsistency Theory and Flying Saucer Sightings," *Science*, 6 Nov. 1970, 599.

19. D. Menzel, *Flying Saucers*, Cambridge, MA: Harvard University Press, 1953.

20. D. Menzel and L. Boyd, *The World of Flying Saucers*, Garden City, NY: Doubleday and Co., 1963.

21. D. Menzel and E. Taves, *The UFO Enigma*, Garden City, NY: Doubleday and Co., 1977.

22. P. Klass, *UFOs Identified*, New York: Random House, 1968.

23. P. Klass, *UFOs Explained*, New York: Random House, 1974.

24. P. Klass, *UFOs: The Public Deceived*, Buffalo, NY: Prometheus Books, 1983.

25. R. Sheaffer, *The UFO Verdict*, Buffalo, NY: Prometheus Books, 1979.

26. J. Oberg, *UFOs and Outer Space Mysteries*, Virginia Beach, VA: Donning Pub. Co., 1982.

27. Lord High Executioner, in *The Mikado* by Gilbert and Sullivan.

28. J. Vallee, *Challenge to Science: The UFO Enigma*, Chicago: Henry Regnery, 1966.

29. C. Sagan and T. Page, ed., *UFOs, A Scientific Debate*, Ithaca, NY: Cornell University Press, 1972.

30. R. Story, *UFOs and the Limits of Science*, New York: Wm. Morrow and Co., 1981.

31. B. Maccabee, "UFOs, the Public Informed," unpublished.

32. Document found in the files of Project *Blue Book*.

33. J. Allen Hynek, *The Hynek UFO Report*, New York: Dell, 1977, 99.

34. D. Menzel, "UFOs, The Modern Myth," see ref. 29, chap. 6.

35. R. Wilson, "A Nuclear Physicist Exposes Flying Saucers," *Look* magazine, Feb. 1951.

36. C. Moore, "Object Report," 24 April 1949 (National Archives microfilm).

37. J. McDonald, written testimony submitted to the Committee on Science and Astronautics of the U.S. House of Representatives of the 90th Congress, in the *Symposium on Unidentified Flying Objects*, July 29, 1968; p. 63.

38. Documents from the Air Force Office of Special Investigations (National Archives microfilm).

39. L. Elterman, "Final Report of Project Twinkle," a report of the Geophysics Research Division of the Air Force Cambridge Research Laboratory, Cambridge, Mass., 29 Nov. 1951 (National Archives microfilm).

40. National Archives microfilm; also *The Mystery of the Green Fireballs*, W. Moore, ed., 1983, available from W. Moore, #247, 4219 W. Olive, Burbank, CA, 91505; also B. Sparks, and J. Clark, "The Southwestern Lights," Parts 1, 2, and 3, *The International UFO Reporter*, 10, nos. 3, 4, 5, (May/June, July/August, and Sept./Oct. 1985), available from the Center for UFO Studies.

41. E. Ruppelt, op. cit., p. 120.

42. Written report by the witness, and private communication; name confidential but available to qualified investigators.

43. Written report by the witness, and private communication; name confidential but available to qualified investigators.

44. W. Startup, *The Kaikoura UFOs*, Auckland: Hodder and Staughton, 1980.

45. Q. Fogarty, *Let's Hope They're Friendly*, Sydney: Angus and Robertson, 1982.

46. B. Maccabee, "Photometric Properties of an Unidentified Bright Object Seen Off the Coast of New Zealand," *Applied Optics* 18, no. 2527 (1979).

47. W. Ireland and M. Andrews, "Photometric Properties of an Unidentified Bright Object Seen Off the Coast of New Zealand," *Applied Optics* 18, no. 3889 (1979).

48. B. Maccabee, "Photometric Properties of an Unidentified Bright Object Seen Off the Coast of New Zealand: Author's Reply to Comments," *Applied Optics* 19, no. 1745 (1980).

49. W. Ireland, "Unfamiliar Observations of Lights in the Night Sky," Report #659 of the Physics and Engineering Laboratory, Department of Scientific and Industrial Research, Lower Hutt, N.Z.

50. Ref. 24, pp. 254 and 255. The author has incorrectly stated that the color of the antiaircraft beacon is red-orange. Actually, the beacon is pure red.

51. The cameraman was very probably looking forward, as were the others, at an unusual flashing light that had appeared. The shape and size of the cockpit would have made it very difficult, and perhaps impossible, for the cameraman to film one of the propellers without displacing the cockpit or the pilot from his seat. No such displacement occurred, however.

52. A study of this effect using color reversal (slide) film under controlled conditions has shown that the outer diameter of the yellowish area is usually as large as, or larger than, the geometric size of the image (i.e., the image size created by simple geometric imaging of the camera). The red fringe extends far beyond the geometric size of the image. The red fringe is created by red light that impinges on the overexposed area and then diffuses sideways within the film. As it diffuses sideways the light intensity decreases and, by the time it has diffused beyond the boundary of the central yellowish area, the intensity of the light is low enough to create proper exposure of the film, that is, a red image. Thus the image of an overexposed red light consists of a pale yellowish center surrounded by a red fringe, the width of which depends upon the intensity of the light.

53. Other explanations that have been offered for this section of the film include (a) a chance alignment of ground level marine navigation beacons, (b) a beacon on another aircraft, (c) a reflection of lights inside the cockpit, (d) a specific beacon in

Wellington Harbor, (e) an astronomical source (not Venus) affected by atmospheric scintillation, (f) a ground-level emergency vehicle and (g) earthquake lights. Each of these has been carefully considered and none has been found to explain the sighting and film.

54. R. Doty, "Alleged Sightings of Unidentified Aerial Lights in Restricted test Range," document File #8017D93-0/29 available from Headquarters, AFOSI, Bolling AFB, Washington, DC 20332, or from the Fund for UFO Research.

55. UFO Fact Sheet, available from the Air Force Office of Public Affairs, Pentagon, Washington, DC.

56. C. Bolender, "Unidentified Flying Objects (UFO)," memorandum dated 29 Oct 1969, available from the Fund for UFO Research.

57. Strategic Air Command Base reports and the NORAD Command Director's Log for the time period 29 October through 10 November 1975; the related documents are available from the Fund for UFO Research.

58. L. Fawcett and B. Greenwood, *Clear Intent*, Englewood Cliffs, NJ: Prentice-Hall, 1984.

59. The information presented here has been extracted from a U.S. Air Force teletype message concerning the debriefing of the second jet pilot, from two in-depth interviews of the air traffic controller and from the interview of the Iranian general who conducted the investigation; available from the Fund for UFO Research.

60. B. Maccabee, "UFO Landings Near Kirtland Air Force Base or Welcome to the Cosmic Watergate," with comments by W. Moore, 1985; available from the Fund for UFO Research.

61. R. Hall, "Sociological Perspectives on UFO Reports," see ref. 29, chap. 9.

62. R. Haines, ed., *UFO Phenomena and the Behavioral Scientist*, Metuchen, N.J.: Scarecrow Press, 1979

63. C. Sagan, *UFOs: The Extraterrestrial and Other Hypotheses*, see ref. 29, chap. 14.

64. I. Shklovskii and C. Sagan, *Intelligent Life in the Universe*, New York: Dell. Pub. Co., 1966; C. Sagan, *The Cosmic Connection*, New York: Dell Pub. Co., 1973.

65. T. Kuiper and M. Morris, "Searching for Extraterrestrial Civilizations," *Science* 196 (6 May 1977), 196.

22

Comment on
Bruce Maccabee

JAMES E. OBERG

The "Null Hypothesis" for UFO reports, of which I am one of a handful of champions, states that no extraordinary stimuli are required to produce the entire array of public UFO perceptions in all their rich variety, wonderment, and terror. Known phenomena have produced all types of what is commonly known as "UFO reports," including apparitions of flying disks, radar and radio interference, terrifying chases, "intelligent maneuvers," telepathic messages, and so forth. There seem to be no types of reports that have not been on record, produced at some point or another by prosaic circumstances.

We can consider the situation as a "black box" that consists of the human sensory/perceptual/mnemonic process. Into one end we insert any of a thousand various types of currency; we turn the crank and activate some undefined algorithm to process the raw stimulus; out the other end comes a "UFO report." We then collect and categorize these reports, and we attempt to define the *inverse algorithm* to recreate the stimulus based on the report. Sometimes we are able to "cheat," to short circuit the black box by going back to records that the stimulus made on other witnesses (different algorithms!) or on other recording media (more hi-fidelity black boxes, with much simpler transformation algorithms). But often the inverse attempt fails.

The argument of the UFO proponents, as represented by Dr. Maccabee, seems to be that since we cannot generate inverse transformational algorithms that convert *some* reports back into recognizably prosaic stimuli, then some truly extraordinary (possibly extraterrestrial and intelligent) stimuli *must exist* to account for our inability to perform the reverse transformation.

I believe this represents a basic fallacy of reasoning and of proof.

Based on evidence as given, the conclusion of "extraordinariness of stimuli" is, in my analysis, *not proven* by any standard of the scientific method.

Perhaps a key misconception of Maccabee's paper, and of pro-UFO argumentation in general, is an inaccurate estimate of the correlation of "simplicity and hi-fidelity of the algorithm" with education and expertise of the percipients. The argument is made that the whiter the collar, the truer the story. That is, percipients with higher education and more technological professions can be counted on to provide accounts that more accurately reflect the actual stimulus. A "trained observer" such as a pilot, being an expert on things seen above the ground, would thus hypothetically be able to provide extremely reliable descriptions of anomalous apparitions. The same argument is made for policemen, for engineers, for astronomers, for other professional people.

But this view is vulnerable to numerous counter-examples (allowing in some cases the generalization that if there is indeed any correlation it is *inverse*, not *direct*), and the view also can be put into question through common sense and everyday experience.

Who, after all, have been the favorite dupes of fake "psychics" such as Uri Geller? It's been the scientists, with their diploma-hung walls, who have time and again been suckered by the crudest sleight-of-hand tricks that wouldn't fool a child (and sometimes it's been children who have fooled the professors). The question here is not intelligence or honesty, it is the very "training" that makes the phrase "trained observer" ring hollow with mocking laughter. Scientists are trained to perceive the real world in a rather particular way, which may or may not accommodate sudden, brief, and unexpected apparitions. A stage magician knows well that he can misdirect and amaze adults (and the more intelligent and imaginative, the better) while often being unable to delude children (since they don't yet have the wealth of experience for readiness to be deliberately and deceptively cued by the performer).

Such cuing and misdirection can also occur accidentally, and when it does, the more intelligent and imaginative percipient is usually at a disadvantage in accurately recounting the stimulus. The "black box" of such a percipient contains an immensely rich and tangled transformational algorithm, making the process of inverting the transformation much more difficult (if not impossible, as examples will show).

Aircraft pilots are an excellent example. Just what does their perceptual training consist of? Maccabee (et al.) would have us believe that decades of cockpit experience have developed in such people a dispassionate reflex to note the characteristics of any and all sudden visual apparitions. But this is unreasonable: surviving pilots are people of rapid action, not calm contemplation. Any unusual perception may instinctively be interpreted immediately in its most dangerous possible incarnation, and avoidance action must be ex-

ecuted quickly. Only later, when a sense of surprise (and by con-
ditioned reflex, danger) has passed, can the pilot react closer to the
human norm, with curiosity and careful observation.

And what is the most dangerous thing a pilot can see out the
window in mid-flight? It is another aircraft, on collision course. The
training in observation that pilots therefore obtain over their flying
careers is to instantly see if the visual apparition is consistent with
some interpretation of an approaching physical aircraft, and if so,
react to avoid collision. "Better safe than sorry" is a prudent motto in
midair.

Hence it should come as no surprise that pilots have repeatedly
misinterpreted distant fire-ball meteors as nearby jets or rockets,
have thrown their aircraft into violent evasive maneuvers to dodge a
falling satellite sixty miles overhead, have made turns to avoid running into the cloud-shrouded rising crescent moon, and similar cases.
Such misperceptions—which err on the side of caution and hence
tend to enhance the survival rate of the percipients—are regular
features of UFO reports by pilots. This is so much so that even Dr. J.
Allen Hynek of the Center for UFO Studies remarked in one of his
books that "Surprisingly, pilots are among the poorest observers of
UFOs"—a valid generalization that Hynek did not draw appropriate
conclusions from, and which he repudiated (or forgot) in later years.

The treatment of the Arnold case was necessarily abbreviated.
Maccabee had to leave out certain features, such as the motion and
shape of the objects (which do not correspond to "classic" flying
saucers), and the "repeater" nature of Arnold's experiences (he has
been quoted as claiming multiple sightings of the same type of UFO
he first saw). Since Arnold's account differed in so many details from
the "run-of-the-mill UFO," perhaps the perception appeared more
characteristic of the internal percipient rather than the external
stimuli. But immediately afterwards, Arnold quickly adopted the
theme of "Flying Disks," a concept practically invented by publicist
Ray Palmer. As a science fiction editor, Palmer knew that circular
spaceships from outer space had been appearing on the lurid covers
of the SF pulps since the mid-1930s if not earlier.

The "trained observer" fallacy appears frequently in Maccabee's
paper, and leads him to attempt to enhance the credibility of certain
percipients by inaccurately exaggerating their professional credentials. A good example is the "Rogue River Case," in which Maccabee
asserts in an appendix to his paper that two "engineers" were the
percipients. The original report actually described the two men as "a
mechanic" and "a draftsman," not as engineers. While a skeptic
would maintain that the men's professions actually had not demon-
strable correlation with their credibility, it is Maccabee's thesis that
the more professional the witness, the more reliable his testimony;
this encourages UFO proponents to exaggerate such professional
status, as Maccabee has done here contrary to the original docu-
mentation of the case. Such an exaggeration is in my judgment

irrelevant to the credibility of the two witnesses in question, but it should perhaps bear some relevance to Maccabee's credibility.

To summarize the point in dispute, I quote Maccabee: "If there were no TRUFOs to be sighted (i.e., TRUFOs don't exist), then this [Battelle Study] result implies that the more reliable (*sic*) a witness is, the more likely he is to make mistakes in his report of a sighting. On the other hand, if there were TRUFO sightings in the sample, then the result is completely consistent with our understanding of reliability." Precisely so—as I have argued that Maccabee's basic understanding of the concept of witness reliability is faulty.

How does one test a "black box"? Analysis of the outputs may provide some statistical insight, but the most practical approach is experimental. One introduces controlled inputs and measures the resulting outputs. Doing so would give an analyst a good appreciation of the nature of the internal transformational algorithm, which maps specific stimuli into a multidimensional array of perceptions, and thus also of the inverse algorithm by which an analyst can attempt to reconstruct the original stimuli based on the reported perceptions.

Staging "fake UFO cases" is one way, but has logistic as well as ethical problems. Better yet would be to find a class of well-defined stimuli and well-documented perceptions, and then define the mapping function. This does not appear to be a line of inquiry followed by pro-UFO specialists, even though it is crucial to their argumentation since it provides otherwise unknowable insights into the issue of "reliability" via an explicit mapping of the behavior of the "black box" in question.

Soviet space and missile activity has provided us with a near-perfect "control experiment" in which startling visual apparitions were seen over a wide expanse of time and space, while the prosaic explanation was not published due to military secrecy.

In 1967, a series of orbit-to-earth thermo-nuclear warhead tests were conducted, with final reentry occurring at dusk across the Ukraine, Volga Valley, and Caucasus regions of the southern USSR. Hundreds of thousands of witnesses stepped forward, including pilots (who saw a "pseudo-UFO" circle their aircraft causing its engines to stall), astronomers (who estimated the solid crescent-shaped "UFOs" to be a thousand feet from wingtip to wingtip), engineers, teachers, chemists, mechanics, and representatives of practically every other respected profession. The were seeing a series of classic "fireball" UFOs, with timing and motion coincident with the descending warheads but *with perceptions identical to those associated with "classic" true UFOs.*

Since 1977, multitudes of witnesses in Argentina, Chile, Uruguay, and southern Brazil have described "circular UFOs" that occasionally fly across the evening sky. Pilots have seen them and chased them; astronomers have dubbed them "classic flying disks"; dockworkers in Chile swore the object rose up dripping from the ocean to the West; motorists have careened down mountain roads in desperate flight;

witnesses have sworn to radar contact and radio interference, and several even testified—under hypnosis and polygraph—to psychic messages and physical sexual contact with occupants of the pseudo-UFOs. All these particular cases can be traced directly and unequivocally to excess propellant dumps from Soviet military missile observation satellites launched an hour earlier from Plesetsk.The visual stimuli were straightforward, but *the black box algorithms were not*, and nobody in the UFO field was able to derive the inverse transform until I matched records of launchings with the times of the mass sightings.

Most recently, in January 1985, the Soviet newspaper *Trud* described an encounter between an airliner and a UFO on an early morning flight from Rostov to Tallinn. The object reportedly paced the airliner, illuminated the ground beneath, was seen by another passing airliner, was observed on radar by ground controllers, and even at one point reflected the image of the airliner. Further research allowed me to correlate this sighting with the launching of an SS-X-25 ICBM from the Plesetsk missile center north of Moscow; the launch was observed from more than a dozen sites in neighboring Finland, about 4 A.M. September 7, 1984. The visual stimulus was the booster exhaust plume; everything else was provided gratis from the imaginations of the involved percipients. *These were air crews and traffic controllers, among the most "reliable" observers according to Maccabee.*

The conclusion is that the "black box" contains transformational algorithms even more bizarre than Maccabee imagines, with all the consequent difficulty in constructing inverses. If prosaic (albeit rare and secret) visual stimuli can thus produce any and all types of known "UFO phenomena," where is the *necessity* to demand the existence of extraordinary stimuli?

Let us see how fruitful the black box approach can be in the analysis of the testability (the "disprovability") of the null hypothesis. If the residue of "true UFO reports" are a function not of extraordinary stimuli but of the investigators' lack of experience and trained insight and the nonavailability of records of candidate prosaic stimui, one would postulate that the earlier in the modern UFO era one looked, the greater the chance of finding so-called "true UFOs." In the first years of the mass phenomenon, investigators had little real understanding of the more arcane aspects of the "black box" transformation and thus they had scant prospect of developing inverses ("solutions") even if they existed; in those years, the number of highly classified flight projects was quite high (e.g., "Skyhook," or later, the U-2) and records of even nonclassified flight traffic were spotty and short-lived. Hence one could predict that any selection of "true UFOs" chosen on a basis of equal "quality" should be heavily biased towards earlier reports.

This is exactly what is seen in Maccabee's selection of twenty-five TRUFOs. An unexpectedly (and nonrandom) large portion are from the first decade of "modern UFOria," and proper weighting for subsequent growth in the pool of potential percipients (general population, pilots and passengers, astronomers, campers, truckers, etc.) would doubtlessly increase the startling asymmetry: there are several times too many "TRUFOs" in the earliest years of this period than one would expect if the phenomenon were independent of the growing experience and insight of investigators. One explanation might be that the "real" stimulus has markedly declined, and another is that the public has stopped bothering to report them. But from the point of view of the null hypothesis, the explanation of this distribution is that investigators have with the passage of time developed a much better knowledge of the "black box" and its functions, both direct and inverse, and thus are solving cases today that would have gone unsolved twenty to thirty years ago. Since investigating cases more than a few years old is a labor fit for Hercules, Hercule Poirot, or a miracle worker—the passage of even a short span of time can render the simplest prosaic "UFO" into a "true UFO" if conditions are right. In general, it seems if not solved quickly a UFO case will probably never be solved (with some notable exceptions).

This skewed distribution has been noted by UFO skeptic Philip J. Klass among others, and is consistent with my observation of the "Mother Goose Syndrome" in UFOlogy: the best stories tend to be "long ago and far away," since these are usually congenitally immune from effective skeptical investigation. As an example, just consider the recent UFOlogical love affair with mainland Chinese UFO reports, despite what seems so obvious (to me) to be connections between major cases and some as yet undocumented Red Chinese missile testing program.

Some shaky analogies in Maccabee's paper cannot be overlooked, either. Treating the UFO phenomenon with the same outlook as the proverb, "Where there's smoke, there's fire," is not a very adept argumentation trick. We all realize that the reliable phrase should simply be, "where there's smoke, there's smoke," since the fallacy of assertion of the consequent (as it is called by logicians) requires that the conclusion ("fire") be the only possible engendering cause of the conditional ("smoke"), when of course common sense and everyday experience tells us it's not (lucky for my kitchen and the motor of my car!).

Another analogy can be conjured up, and a syllogism constructed to test it. A leading pro-UFO argument is that since some UFO reports cannot be solved, then their stimulus must be extraordinary. Isn't this akin to suggesting that since some missing children are never found, they might well be on Mars? It turns out there really is a "Judge Crater" on the moon, for example. Since some airplanes and automobiles crash without explanation, are extraterrestrial traffic saboteurs at work? Since a good fraction of murders remain forever unsolved, are psychotic time-traveling killer-robots at work? Surely not, of course—it's absurd even to suggest such ideas. But how far afield is the analogy to the evidential value of "unsolved" UFO cases?

The syllogism, to repeat, goes like this: since some UFO reports cannot be solved by amateur investigators working in their spare time, then there must exist extraordinary stimuli behind some UFO reports. As a technique of analysis, invert the syllogism to create a new one of equivalent Boolean value. It now reads: if all UFO reports were caused only by ordinary stimuli, then amateur investigators working in their spare time would be able to solve every one of them. Worded this way, the syllogism is arguably untrue (there are numerous counterexamples of cases that happened to be solved only by "accident"); its untruth implies untruth for the first equivalent assertion, championed by UFO proponents.

What then do I as a skeptic conclude about prospects and policies for the UFO phenomenon?

The possibility nevertheless still exists that currently undefined phenomena of genuine interest to science (or, say, to the CIA) are occurring but are being perceived and reported as "UFOs," and hence are not coming to the attention of the proper specialists. For that reason, serious continuing efforts are justified both by Maccabee and his colleagues and by the skeptics. I encourage donations to the Fund for UFO Studies.

Just because a perception goes into the black box and comes out transformed into a "UFO report" is no grounds whatsoever for concluding that the original stimulus was uninteresting. As is well known in mathematics, a mapping function can be many-to-one (that is, different inputs can produce the same outputs) and experience suggests that our UFOric black box does behave that way, making any unambiguous reverse transformation mathematically impossible. So "extraordinary stimuli" are *not disproved* by this line of reasoning, which can only claim that such stimuli are *not proved*.

23

Response to James Oberg's Comment Paper

BRUCE MACCABEE

Mr. Oberg has stated that he is a champion of the "null hypothesis," which is that "Known phenomena have produced all types of what is commonly known as 'UFO Reports.'" Furthermore, according to Oberg, "no extraordinary stimuli are required to produce the entire array of public UFO perceptions." In other words, Oberg claims that no UFO reports have been caused by sightings of TRUFOS, as defined in my paper (phenomena not yet recognized as real by the community of scientists). Considering the high quality of the sightings, in terms of descriptive information and credibility of the witnesses, Mr. Oberg's claim cannot be accepted without proof.

He has tried to support his position by emphasizing the fallibility of witnesses. His argument is well written but, like a clever magician, he has diverted the reader's attention from the central issue, which is to *explain each sighting*. Mr. Oberg has not actually explained any of the sightings listed in my paper because the "black box" hypothesis that he claims explains all UFO reports *does not actually apply to the sightings I have presented*.

He argues that the phenomenon that caused a sighting cannot be determined with any degree of confidence by analyzing the *interpretation* given by the witness(es) or by news media reports of the sighting (his "black box" hypothesis). I (and other scientific UFOlogists) agree with this claim. Therefore I have analyzed (and discussed in my paper) the *descriptions* of the phenomena as given by the witnesses rather then their *interpretations* of what they saw.

To support his argument that the witness' *interpretation* of what he (she) saw is very often incorrect, Mr. Oberg has cited the example of the aircraft pilots who "have repeatedly misinterpreted distant fireball meteors as nearby jets or rockets, have . . . (dodged) a falling satellite sixty miles overhead, have made turns to avoid running into the cloud-shrouded rising crescent moon" and so on. To further support his argument, he has discussed sightings of "circular UFOs" in South America. Mr. Oberg pointed out that the witnesses and the news media often interpreted these phenomena as alien spacecraft ("UFOs") even though they were caused by Russian rocket launches and propellant dumps. Unfortunately, Mr. Oberg has written his paper in such a way that the reader (and perhaps Mr. Oberg as well) might conclude that, because the interpretations were wrong, the *descriptions* were also wrong. However, the fact is that the descriptions were quite often very accurate.

Mr. Oberg was the first UFOlogist to try to connect sightings in Russia and South America with Russian rocket launches. In trying to make this connection he was confronted with extraordinary descriptions by the witnesses. The witnesses described seeing glowing semi-transparent "things" at very high altitudes. The "things" had "tentacles" hanging downward (e.g., the Petrazavodsk "jellyfish UFO" of Sept. 1977). Similar sightings occurred in South America. It is a tribute to Mr. Oberg's diligence in the field of UFOlogy that he managed to identify the causes of these sightings.

Mr. Oberg very admirably demonstrated the ability of witnesses to give accurate descriptions of unusual phenomena. But in his paper Mr. Oberg used the fact that these sightings were widely interpreted as sightings of alien craft to demonstrate the untrustworthiness of witness and tabloid *interpretations* and therefore, by inference, the untrustworthiness of witness descriptions. He did not mention, in his written discussion, that descriptions given by the witnesses were sufficiently accurate to allow Mr. Oberg to claim that he has *positively identified the phenomenon.* He demonstrated the accuracy of the witnesses during his ICUS meeting lecture when he showed first some drawings by witnesses and then some photographs of the phenomenon. The comparison was striking; the witness drawings had all the major details that were evident in the photographs. The comparison of descriptions and drawings with the photographs leaves no room for doubt: Mr. Oberg is correct in claiming that these were *not* sightings of alien craft, but rather of Russian rocket launches.

Thus Mr. Oberg demonstrated what has been known to UFOlogists (and criminal investigators) for many years, namely that once the *interpretations* are stripped away from the reports *the remaining "pure" descriptions are generally quite accurate.* Furthermore, *better trained observers give more precise descriptions* although, as Mr. Oberg pointed out, they may be just as wrong as any other observers in their *interpretations* of what they saw.

Although it is true that the accuracy of witness descriptions is lower for transient phenomena (e.g., meteors) and for very complex phenomena (multiple "objects" or phenomena moving in various complicated ways), even these descriptions are generally good enough to establish a probable identification. Keeping this in mind, the reader is invited to reread the discussions of the relatively *long duration and uncomplicated* sightings described there (e.g., Arnold saw the objects for over two minutes, the Rogue River fishermen/engineers saw the object for over two minutes, etc.) (Note: a much better feeling for the accuracy of descriptions can be obtained by reading the *original* government documents.) It is true that Arnold's interpretation was wrong: he thought he saw a new type of high-speed jet aircraft. This interpretation did not affect his description: he did not report wings or tails on the objects, despite his attempt (wish?) to see them.

Finally, it should be noted that *most UFO reports are explained by UFOlogists precisely because the witness' descriptions are accurate enough to allow for a reasonably confident identification of the phenomenon.*

The following are brief comments on some of the other points in Oberg's paper:

1. Oberg's "black box" argument does not apply to film. I listed (Table 21-1) several incidents involving movie film or still photographs (of course there are many more), none of which were discussed by Mr. Oberg. It is also difficult to apply his argument to instrumented sightings such as were made by the General Mills high altitude balloon project personnel (e.g., C. B. Moore, Apr. 24, 1949) or by other meteorologists.

2. Mr. Oberg has criticized my discussion of Kenneth Arnold's sighting (June 24, 1947) for not describing the "motion and shape" of the objects. In response to this comment I have included, in the published version, a few more details about the objects (e.g., they were crescent shaped and tilted back and forth as they flew).

His comment about them not resembling "classic" saucers is irrelevant. It should be noted that the term "flying saucer" was coined by a newsman who evidently tried to compress (as the newsmedia always do) two elements of Arnold's description: crescent-shaped *flying* objects, and tilting while flying, like *saucers* skipping across the water. Oddly enough, some of the first photographs (taken in Phoenix, Arizona, in July 1947) show *crescent-shaped objects.*

Mr. Oberg also referred to Arnold as a "repeater." This is an attempt to diminish Arnold's credibility (because no one should be able to see something rare . . . or nonexistent . . . more than once; therefore, the sighting is either a misidentification or a hoax). However, this is another irrelevant claim since this was Arnold's *first* sighting.

3. Mr. Oberg has criticized me for referring to two of the Rogue River witnesses as "engineers," when, in fact, one was a mechanic and one was a draftsman. (O.K., I'll buy that.) However, he has somehow

overlooked a much more important facet of the sighting: the object that was reported to the Security Force of the Ames Research Laboratory by the mechanic and the draftsman (and also seen by three other people) *has not been identified and, in fact, would seem to be unidentifiable.*

Mr. Oberg has tried to impugn my credibility by implying that I chose the word "engineers" to make the reader think that the men were more qualified than they were, but this is false. On the other hand, Mr. Oberg's treatment of this sighting puts his credibility in question because, although he has claimed that all sightings can be explained, *he has not made any suggestion as to what the witnesses saw if it weren't a TRUFO!*

4. Mr. Oberg has tried to make the conclusion of the Battelle Institute study, that the more reliable the observer, the *less likely is it that a sighting can be explained,* appear to agree with his "trained observer fallacy" hypothesis. However, the examples that he provided in his prepared lecture (descriptions of Russian rocket launches and propellant dumps misinterpreted as vehicles from outer space) *are completely consistent with the idea that better observers provide more accurate descriptions.*

5. Mr. Oberg has claimed that I purposely selected early sightings (e.g., pre-1960) because the "investigators' lack of experience and trained insight and the nonavailability of records of candidate prosaic stimuli" would increase the "chance of finding so-called 'true UFOs.'" However, this is patently wrong. I listed the reasons for my choices. I could have concentrated on sightings made *since* Project Blue Book closed (1969), for example, but instead I chose to include important "historical" sightings to demonstrate that *evidence to establish the reality of TRUFOs has been available for many years and it is only the reluctance of the scientific community to come to grips with the* "hard core" sightings that has kept this problem from being solved.

Furthermore, contrary to what Oberg claims, many of the earlier sightings *are better reported and better investigated than more recent ones.* Mr. Oberg is weak in his history if he thinks, for example, that Project Blue Book investigations improved with time after 1952 and that this accounts for the drop in the percentage of unknowns after 1953. The actual fact is that Blue Book investigations of sightings became poorer after 1953 and the drop in the percentage of unknowns resulted largely from bookkeeping practices that "converted" sightings initially listed as "possibly known" or "insufficient information" into "known" at the end of a year (see, e.g., Ref. 3 of my paper). This "conversion" increases the number of "knowns" relative to the unexplained sightings and thus reduced the percentage of unknowns to about 3 percent of the total reports, a percentage which remained about constant for many years.

In conclusion, Mr. Oberg's paper has not solved the UFO problem.

In fact, his work has added credibility to the problem, since he has established the accuracy of witnesses when confronted with a completely baffling and unknown phenomenon (rocket launches and fuel dumps).

Any reader wanting further information on the sightings listed in Table 21-1 is invited to write to this author at the following address: Fund for UFO Research, Box 277, Mt. Rainier, MD 20712.

24

Comment on
Bruce Maccabee

WILLIAM H. STIEBING, JR.

In recent years a number of authors have claimed that UFOs are not a recent phenomenon—that earth was visited by individuals from space in very early times. These advanced visitors from the skies became the gods and heroes of mankind's myths, and they not only helped create civilization on this planet, but they also may have been responsible for the development of homo sapiens from primitive primates. While much of the evidence for this thesis is drawn from subjective interpretations of myths, the most persuasive arguments rely on inferences drawn from the existence of archaeological "mysteries" in various parts of the world. However, archaeological evidence does not really support the theory that earth received visitors from space in antiquity. The archaeological material has been misunderstood and/or misused by the supporters of the ancient astronaut thesis.

Bruce Maccabee's discussion of UFOs naturally concentrates on sightings since 1947. But, as he remarks at the beginning of his paper, unusual flying objects occasionally have been reported since ancient times. Are UFOs a modern phenomenon or has earth been receiving visitors from space for thousands of years? Most of the ancient accounts of strange objects in the sky occur in myths or religious visions, so scholars have not usually treated them as references to objective physical phenomena. However, in recent years, a number of writers have taken these early accounts more seriously, using them as evidence that astronauts from outer space created human beings by genetic manipulation or interbreeding, then taught the new creatures the basics of civilization.[1]

Velikovsky used myths and legends to support a theory of cosmic catastrophes by assuming that they reported real observations that the ancients did not know how to describe in scientific terms. Those who use mythology as evidence that earth was visited by spacemen in antiquity made the same assumption. However, as Raphael Patai pointed out in "Myths of the Universe: Cosmogony and Cosmology" (this volume) myths are not just primitive descriptions of experience or an inexact form of history. Myths validate customs, laws, rites, or religious beliefs, as well as offer psychologically satisfying explanations for man's relationships with his fellow men and with the world around him. Attempts to read them as fairly literal accounts of natural events, whether those events be comets colliding with earth or astronauts coming down from the skies, are very subjective and impressionistic. This is not really very suitable evidence for a scientific or historical thesis.

More persuasive than interpretations of myths or biblical passages are the archaeological "mysteries" that are cited as circumstantial evidence for ancient visitors from outer space. It is this evidence that was featured most prominently in two popular movies and two television specials presenting the ancient astronaut thesis, and it seems to be the evidence that is most convincing to laymen.

For example, a map belonging to the sixteenth century Turkish admiral Piri Re'is supposedly depicts with amazing accuracy the coasts of North and South America as well as land that might be Antarctica:

> The mountain ranges in the Antarctic, which already figure on Reis' maps, were not discovered until 1952. They have been covered in ice for thousands of years, and our present-day maps have been drawn with the aid of echo-sounding apparatus. . . . Comparison with modern photographs of our globe taken from satellites showed that the originals of Piri Reis' maps must have been aerial photographs taken from a great height.[2]

In addition, there are gigantic lines etched into the surface of the coastal plain of Peru near Nazca that have the appearance of an airfield.[3] And in various parts of the world there are representations of individuals in what appear to be spacesuits with helmets and other accessories.[4] It is even claimed that a Mayan sculpture from Palenque shows a man inside a spaceship that is taking off.[5] Finally, it is asserted that the Great Pyramid of Giza in Egypt, Stonehenge in England, the gigantic statues on Easter Island in the Pacific, the huge "Gateway of the Sun" at Tiahuanaco in Peru, and other megalithic monuments could not have been created with the limited technology of early cultures. These huge stone artifacts are cited as indirect testimony that in ancient times earth was visited by individuals possessing a very advanced technology.[6]

While Erich von Däniken and others make their case look very convincing to the layman, the archaeological evidence does not really support their claims. If one studies the Piri Re'is map very carefully, it will become evident that claims about its amazing accuracy are false. Cuba is incorrectly labeled "Hispaniola" and is drawn totally out of proportion. The Virgin Islands are in the wrong positions, incorrectly shaped and badly out of scale. Nine hundred miles of South America's coastline are missing, the Amazon River is shown twice and there is no sea passage indicated between South America and "Antarctica." Finally, despite claims to the contrary, the coast of "Antarctica" on the map bears little resemblance to either the present-day coastline or seismic profiles of the land below the ice cap. The explanatory notes on the "Antarctica" section of the map state that the land is inhabited by white-haired monsters, six-horned oxen, and large snakes, and credits this information to Portugese sailors and their maps.[7] Neither spacemen nor photographs from orbiting satellites are needed to explain the map. It seems to be exactly what Piri Re'is claimed it was in a comment written in the margin—a composite map based on older Greek and Arab charts, on a map made by Columbus, and on charts as well as verbal reports of other early New World explorers. It is an excellent example of sixteenth century mapmaking, but it is a *human* product, not a "mystery" derived from ancient space visitors.

Unlike the Piri Re'is map, the Nazca lines really are a "mystery." No one knows why the large animal figures and straight lines which run for miles were drawn on the plain at Nazca. But the work was not directed from hovering spacecraft or planned as landing strips. The surface of the Nazca plain is relatively soft—no spaceship would ever be able to land or take off there. In addition, small scale drawings have been found that indicate that the natives of the area made the large drawings by using a grid to enlarge the small sketches. The presence of Nazca pottery shards all over the plain and the similarity between animal figures on the pottery and those drawn on the plain indicate that the lines were almost certainly created by the Nazca culture which flourished between c. 200 B.C. and A.D. 600. Since these people worshipped the sun, moon, and other heavenly bodies, it is not so strange that they spent a great amount of time and energy making figures that could be properly seen and appreciated only from the air.[8]

The interpretations of rock carvings in Algeria, Australia, Palenque and elsewhere as representations of spacemen with spacesuits, helmets, and antennae are also not convincing when the drawings are studied in detail and in context (including related carvings that are not illustrated by von Däniken and his supporters). Many of the figures, for example, are clearly naked except for the large round "helmets" they wear (breasts, fingers, and toes are shown). Furthermore, these "advanced space visitors" often carry bows and arrows rather than the laser guns or other advanced weapons one would expect. The "helmets" are probably ceremonial masks worn during religious dances. Not only is the Palenque "astronaut" nude except for a loin cloth and some jewelry, but his "spaceship" is a stylized representation of a maize plant or "tree of life." Other sculptures at the same site make this very clear, but they are not usually illustrated in the books promoting the ancient astronaut thesis.[9]

The pyramids, Easter Island statues, and other huge monuments are also the products of *human* technology and creativity. The claim of von Däniken and his supporters is that primitive cultures *could not* have constructed such gigantic structures. Yet Egyptian tomb reliefs show stones weighing as much as sixty tons being moved on a sled pulled by a gang of 172 men. And Egyptian quarries produce evidence of how the stones were split from the rock face using very primitive tools. Furthermore, the evolution of the design of the Great Pyramid from earlier mastaba tombs and step pyramids and the evidence of changes in its design during construction show that it was not planned as a monumental reference work whose measurements contained secrets of the universe. Study of other monuments has produced similar results. The natives of Easter Island have demonstrated for modern researchers the traditional methods for carving, moving, and erecting their huge statues. Like the ancient Egyptians, they used nothing more than simple tools, wooden levers, sleds, and ramps. And a Jesuit chronicler recorded that the Incas of Peru were still constructing their monumental buildings in much the same way in the seventeenth century. No one explained to these "primitive" peoples that it was impossible for them to construct their monuments with only their simple technology. Only by ignoring the evidence of archaeology and creating such supposed "impossibilities" can it be claimed that the inspiration and know-how behind these and other "mysteries" came from outer space.[10]

The most perverse aspect of the space visitor theory is that it denies to human beings many of their greatest accomplishments and most amazing creations. It opts for the simple solution to all questions about origins—everything new and worthwhile was the work of spacemen. But the evidence indicates that it was human ingenuity working independently on similar problems that figured out how to move the stones for the pyramids or Stonehenge or Easter Island. Whether or not there are modern UFOs is still an open question. But the situation in regard to antiquity is clear—there is no valid evidence for visits by ancient astronauts.

The consideration of the ancient spacemen theory brings us in a full circle to where this symposium began—to a consideration of myth. For pseudoscientific theories like those of von Däniken are the new mythology for the modern era. They function in the same way that myths always have, resolving psychological dilemmas and providing simple answers for the unknown or unknowable. Science cannot explain everything or answer all questions. It is especially unable to answer questions dealing with purpose, meaning, or ultimate origins. But many people have a low tolerance for the psychological discomfort caused by unresolved problems or unanswered questions. They need answers, so they turn to myth. Traditional myths still meet these needs for some, but others desire myths disguised as science, "theories" with archaeological, textual, and scientific "evidence" supporting them. The "evidence" need not be strong. It only has to be convincing enough to reassure those who already want to believe. Just as science will never totally replace or extinguish mythology, so there will always be some individuals who prefer pseudoscience to science. However, it is important that the public constantly be reminded of the differences between these divergent ways of understanding the universe around us.

NOTES

1. The best known work is E. von Däniken's *Chariots of the Gods?* New York: Bantam Books, 1968; but see also, for example, A. Tomas, *We Are Not the First*. New York: Bantam Books, 1971; J. Bergier, *Extraterrestrial Visitations from Prehistoric Times to the Present*. New York: Signet Books, 1974; J. Blumrich, *The Spaceships of Ezekiel*. New York: Bantam Books, 1974, A. and S. Landsburg, *In Search of Ancient Mysteries*. New York: Bantam Books, 1974; Z. Sitchin, *The Twelfth Planet*. New York: Avon Books, 1976; or M. Chatelain, *Our Ancestors Came from Outer Space*. London: Arthur Barker, 1977. For a more detailed analysis of this theory than can be offered here, see R. Story, *The Space-Gods Revealed: A Close Look at the Theories of Erich von Däniken*. New York: Harper & Row, 1976; or W. Stiebing, Jr., *Ancient Astronauts, Cosmic Collisions, and Other Popular Theories About Man's Past* (Buffalo, New York: Prometheus Books, 1984) 81–106.
2. Von Däniken, *Chariots of the Gods?*, 15. See also Tomas, *We Are Not the First*, 87–90.
3. Von Däniken, *Chariots of the Gods?*, 16–18; Landsberg, *In Search of Ancient Mysteries*, 57–69.
4. Von Däniken, *Chariots of the Gods?*, 31–33; Tomas, *We Are Not the First*, 117–118.
5. Von Däniken, *Chariots of the Gods?*, 110–111; Tomas, *We Are Not the First*, plates 12 and 13 in the photo section in the center of the book; Chatelain, *Our Ancestors Came from Outer Space*, 63.
6. Von Däniken, *Chariots of the Gods?*, 19–22, 74–79, 90–91; Landsberg, *In Search of Ancient Mysteries*, 99–119; Chatelain, *Our Ancestors Came from Outer Space*, 65–83.
7. See C. Hapgood, *Maps of the Ancient Sea Kings: Evidence of Advanced Civilization in the Ice Age*. (New York: Chilton Books, 1966) 60–77 and Fig. 18 and story; *The Space-Gods Revealed*, 29–31.
8. Stiebing, Jr., *Ancient Astronauts, Cosmic Collisions*, 94–96.
9. *Ibid.*, 98–101.
10. *Ibid.*, 101–121.

REFERENCES

Bergier, J. *Extraterrestrial Visitations from Prehistoric Times to the Present*. New York: Signet Books, 1974.
Blumrich, J. *The Spaceships of Ezekiel*. New York: Bantam Books, 1974.
Chatelain, M. *Our Ancestors Came from Outer Space*. London: Arthur Barker, 1977.
Hapgood, C. *Maps of the Ancient Sea Kings: Evidence of Advanced Civilization in the Ice Age*. New York: Chilton Books, 1966.

Landsburg, A. and S. Landsburg, *In Search of Ancient Mysteries*. New York: Bantam Books, 1974.

Sitchin, Z. *The Twelfth Planet*. New York: Avon Books, 1976.

Stiebing, Jr., W. *Ancient Astronauts, Cosmic Collisions, and Other Popular Theories About Man's Past*. Buffalo, New York: Prometheus Books, 1984.

Story, R. *The Space-Gods Revealed: A Close Look at the Theories of Erich von Däniken*. New York: Harper & Row, 1976.

Von Däniken, E. *Chariots of the Gods?* New York: Bantam Books, 1968.

PART **X**

EPILOGUE

25

Science Fiction and Faith

BEN BOVA

Knowledge is always based on faith. The verb *to know* depends on the verb *to believe*.

In the misty beginnings of human society, knowledge and faith were so closely intertwined that the tribal wise man was the tribal shaman. Even in sophisticated societies such as those of ancient Egypt, China, and the Middle East, the men who watched the stars and predicted the seasons were priests, servants of the gods who they fully believed were responsible for the stars and their motions in the sky, as well as the seasons here on earth.

In 1620, the year that the pilgrims set foot on Plymouth Rock, Francis Bacon published in England a book that changed the world. It was titled *Novum Organum*, meaning "The New Method." This book signaled the beginning of a new era in human thought—the organized and self-checking system of observing and measuring that we now call *the scientific method.*

"Man is the servant and interpreter of Nature," wrote Bacon. "About Nature, consult Nature herself."

Early in the seventeenth century a detailed investigation of nature was a new and daring idea. Most learned men of that era were content to discuss "natural philosophy," as physical science was called then, in much the same way that they discussed politics: from the safety of a comfortable chair. There were dangers involved in examining nature. In Italy, Galileo was haled before the Inquisition and kept under house arrest for the remainder of his life because he espoused scientific investigation rather than bow to the authority of the Church.

Bacon titled his book *Novum Organum* in reaction to Aristotle's

269

Organon, which had been written some twenty centuries earlier. During the intervening eighty human generations, Aristotle's word had been the final authority on nearly every question concerning the natural world, throughout Christian Europe.

The flowering of modern science broke the authority of Aristotle and the Church. In ways that most historians still do not appreciate, it was the scientific revolution of the seventeenth century that made possible—even inevitable—the political revolutions that followed.

Yet the scientific method, for all its rationalism and insistence on unbiased measurement and experiment, is based on faith. Not necessarily on a belief in a deity, but on the faith that nature is absolutely indifferent to human actions. The scientist believes, without even thinking consciously about it, that one and one *always* add up to two, that hydrogen atoms behave the same way in distant stars as they do in earthly laboratories, that the universe is predictable—no matter who is making the measurement, no matter what his religion or skin color or politics. The universe plays fair; it is governed by a set of rules that human beings can discover and understand.

Albert Einstein expressed that faith in these words; "The eternal mystery of the world is its comprehensibility."

It is no coincidence that the first science fiction stories—tales that dealt with the wonders of science and exploration of other worlds in space—were written in the seventeenth century. In 1657 Cyrano de Bergerac (the real one, not the romantic figure of Rostand's nineteenth-century play) wrote his *Voyages to the Sun and the Moon* in which he "invented" the idea of using rockets to travel to other worlds.

Fantastic tales had been a mainstay of the world's literature since long before the invention of writing. But unlike these earlier stories, which depended on the supernatural interventions of gods and goddesses, the new "scientific romances" based their adventures on what was known of the real world—plus some informed speculation about what might be discovered in the future.

Like the scientists themselves, the science fiction writers have been guided by a basic tenet of faith: the belief that the universe is knowable, and that what human beings can understand, they can eventually learn to control.

Not everyone shares this belief. Since Galileo's time, and right up to this day, many people have feared science and scientists. Since the Neolithic tribal shamans first arose, the power of knowledge has frightened the ignorant. Most people have held an ambivalent attitude toward the shaman-astrologer-wizard-scientist. On the one hand they envied his abilities and sought to use his power for their own gain. On the other hand, they feared that power, hated his seeming superiority, and knew that he was in league with the dark forces of evil. Many still fear the so-called "Faustian bargain," the mistaken idea that knowledge can only be bought at the price of eternal damnation—or its modern equivalent, nuclear holocaust.

There are two kinds of people in the world: Luddites and Prom-etheans—those who fear science and its offspring technologies, and those who embrace them.

History has not been kind to Ned Ludd, the unwitting founder of the Luddite movement of the early nineteenth century. Webster's *New World Dictionary* describes Ludd as feeble-minded. The *Encyclopaedia Britannica* says he was probably mythical.

The Luddites were very real, however. They were English crafts-men who tried to stop the young Industrial Revolution by destroying the textile mills that were taking away their jobs. Starting in 1811, the Luddites rioted, wrecked factories, and even killed at least one employer who had ordered his guards to shoot at a band of rioting workmen. After five years of such violence, the British government took harsh steps to suppress the Luddites, hanging dozens and trans-porting others to prison colonies in far-off Australia. That broke the back of the movement, but did not put an end to the underlying causes that had created the movement. Slowly, painfully, over many generations, the original Luddite violence evolved into more peaceful political and legal activities. The labor movement grew out of the ashes of the Luddites' terror. Marxism arose in reaction to capitalist exploitation of workers. The Labour Party in Britain, and socialist governments elsewhere in the world, are the descendants of that early resistance against machinery.

Today the progeny of those angry craftsmen live in greater com-fort and wealth than their embattled forebears could have dreamed in their wildest fantasies. Not because employers and factory owners suddenly turned beneficent. Not because the labor movement and socialist governments have eliminated human greed and selfishness. But because the machines—the machines that the Luddites feared and tried to destroy—have generated enough wealth to give common laborers houses of their own, plentiful food, excellent medical care, education for their children, personally owned automobiles, televi-sion sets, refrigerators, stereos, all the accouterments of modern life that we take so much for granted that we almost disdain them, but which would have seemed miracles beyond imagination to the orig-inal Luddites.

We still have the Luddite mentality with us today: people who distrust or even fear the machines that we use to create wealth for ourselves. The modern Luddites are most conspicuous in their resist-ance to high technology such as computers and automated ma-chinery, nuclear reactors, high-voltage power lines, airports, fertil-izers, and food additives. To today's Luddites, any program involving high technology is under immediate and intense suspicion. In their view, technology is either dangerous or evil- or both, and must be stopped. Their automatic response is negative; their most often-used word is *no*.

Opposing the Luddite point of view stands a group of people who fear neither technology nor the future. Instead, they rush forward to try to build tomorrow. They are the Prometheans, named so after the demigod of Greek legend who gave humankind the gift of fire.

Every human culture throughout history has created a Prometheus myth, a legend that goes back to the very beginnings of human consciousness. In this legend, the first humans are poor, weak, starving, freezing creatures, little better than the animals of the forest. A godling—Prometheus to the Greeks, Loki to the Norse, Coyote to the Plains Indians of North America—takes pity on the miserable humans and brings down from the heavens the gift of fire. The other gods are furious, because they fear that with fire the humans will exceed the gods themselves in power. So they punish the gift-giver, eternally.

And, sure enough, with fire the human race does indeed become the master of the world.

The myth is fantastic in detail, yet absolutely correct in spirit. Fire was indeed a gift from the sky. Undoubtedly a bolt of lighting set a tree or bush ablaze, and an especially curious or courageous member of our ancestors overcame the very natural fear of the flames to reach out for the bright warm energy. No telling how many times our ancestors got nothing for their troubles except burned fingers and yowls of pain. But eventually they learned to handle fire safely, to use it. And with fire, technology became the main force in human development.

Technology is our way of dealing with the world, our path for survival. We do not grow wings like the eagle, or fur like the bear, or fleet running legs like the deer. We make tools. We build planes, we make clothing, we manufacture automobiles. The English biologist J. B. S. Haldane said, "The chemical or physical inventor is always a Prometheus. There is no great invention, from fire to flying, that has not been hailed as an insult to some god."

In a broader context, we might say that the basic difference between the Luddites and the Prometheans is the difference between an optimist and a pessimist. Is the glass half full of water or half empty?

Many human beings see themselves, and the entire human race, through the weary eyes of ancient pessimism. They see humankind as a race of failed angels, inherently flawed, destined for eternal frustration. Thus we get the myth of Sisyphus, whose punishment in Hades was to struggle eternally to roll a huge stone up a hill, only to have it always roll back down again as soon as he got it to the summit. It sounds very intellectual to be a pessimist, to adopt a world-weary attitude; at the very least, no one can accuse you of enthusiasm or youthful naïveté.

The optimists tend to see the human race as a species evolving toward immortality. We are perfectible creatures, they believe. Optimists can be accused of naïveté, but they can also point to recent history and show that human thought has improved the human condition immensely within a few short centuries in which science has come into play. Certainly there are shortcomings, pitfalls, drawbacks to every advance the human race makes. But the optimists look to the future with the confidence that humankind can use its brains, its hands, and its heart to constantly improve the world.

It is this difference between the pessimists and optimists that causes a fundamental resistance among the pessimists to science fiction. Especially among those who specialize in the literature of the past, the optimistic literature of a brighter tomorrow is anathema. They simply cannot fathom it; they are blind to what science fiction says. Even *within* the science fiction field, its practitioners often fall prey to this ancient schism, regarding darkly pessimistic stories as somehow more "literary" than brightly optimistic ones.

There is a bit of the pessimistic Luddite in each of us, and each of us is something of an optimistic Promethean.

The great poet and historian, Robert Graves, revealed a decidedly Luddite side of his personality some twenty years ago when he wrote in the British journal *New Scientist:*

"Technology is now warring openly against the crafts, and science covertly against poetry."

Graves apparently feared that the machines that had replaced human muscle power and handiwork—the machines that the Luddites tried to wreck two centuries ago—are now giving way to electronic systems that threaten to replace human brainpower.

He went on to contrast science with poetry, stating that poetry has a power that scientists cannot recognize because poetry is usually the product of intuitive thinking, which "scientists would dismiss . . . as 'illogical.'"

Graves apparently pictured scientists as sober, plodding, soulless thinking machines that do everything rationally and never take a step that has not been carefully examined beforehand. As a historian, he should have known better.

Because scientists have an inherent faith in the "fairness" of the universe (one and one always equals two) they frequently feel free to make leaps into the unknown, confident that they will not find themselves in a region where the basic rules of the universe have changed.

James Clerk Maxwell's brilliant insight that visible light is a form of electromagnetic energy; Max Planck's notion that all forms of energy come in discrete packages, or *quanta;* Wolfgang Pauli's faith in the conservation of energy, which led him to predict the existence of an unseen particle, the neutrino—the history of science is dotted with examples of "leaps of faith."

Scientists are just as human, just as intuitive, just as emotional as poets. They have just as strong a need for a belief in the basic justness of the universe as any monk or minister. But the scientist's belief is in a rational, understandable, predictable universe, rather than a universe created and governed by a personality.

The science fiction writer goes one step farther. Not only do the writers share the scientists' belief in an understandable universe, the writers also tacitly believe that human-like *intelligence* is a permanent fixture in the universe.

Even in the darkest, most dystopian science fiction stories, intelligent life somehow endures. Despite nuclear holocaust, environmental disaster, devastating invasions from other worlds, science fiction stories always show that life goes on—intelligent life persists, somewhere, somehow.

The intelligent life may not be human. In some stories the human race may perish, but that is not the end of the universe. There can be intelligent creatures from another world, or intelligent machines left behind by their creatures who had died away or vanished.

In actuality, what the science fiction writers are doing is creating a modern mythology. Science fiction is slowly, almost unconsciously, codifying a system of beliefs tailored to our modern scientific society.

Joseph Campbell, late Professor of Literature at Sarah Lawrence College, spent his long lifetime studying and writing about mythology. His *Hero with a Thousand Faces* and his four-volume *The Masks of God* are classic books on the world's mythologies.

Campbell has pointed out that modern scientific society has no mythology of its own, no psychological underpinning. The old myths are dead, and no new mythology has arisen to take its place.

A human being needs mythology, he insists, to give an emotional meaning to the world in which we live. A mythology is a codification on the emotional level of our attitudes toward life, death, and the entire vast and often frightening universe.

The Prometheus myth is an example of what Campbell is driving at. It explains, at the deepest emotional level, not merely our mastery of fire and its resulting technology, but the price that must be paid for such a gift.

Much of today's emotion-charged, slightly irrational urge toward astrology and mysticism is actually nothing more than a groping for a new mythology, a set of beliefs that can explain the modern world on the emotional, intuitive level to people who are frightened that they are too small and too weak to cope with society.

Campbell's work has shown that there are at least four functions that a mythology must accomplish, if it is to be helpful to the individual.

First: a mythology must induce a feeling of awe and majesty in the people. Science fiction stories often do this, especially when they strike the reader with new and stunning visions of the universe. In the science fiction community, this is called "a sense of wonder," and it is something that writers strive to achieve.

Second: a mythology must define and uphold a self-consistent system of the universe, a pattern of believable explanation for the phenomena of the world around us. Virtually every science fiction tale takes as a "given" the universe revealed by scientific investigation. While most science fiction writers feel free to extrapolate from what is known today in an effort to produce fresh insights, the writers almost always try to stay within the confines of what is known to be possible. The rule of thumb is: the writer is free to invent anything, as long as it cannot be *proven* to be physically impossible by today's knowledge.

Third: a mythology usually supports a specific social establishment. For example, the body of mythology that originated in ancient Greece apparently stemmed from the Achaean conquerors of the earlier Mycenaean civilization. Zeus was a barbarian sky god who conquered the local goddesses as the Mycenaean cities fell to the invading Achaeans. Many a lovely legend was started this way. Science fiction stories tend to support the social system we call Western Civilization. The concept that the individual person is worth more than the Organization—even when the Organization might be an all-conquering interstellar empire—is one of the basic tenets of science fiction. Especially in Western science fiction, nothing is as important as human freedom.

Fourth: a mythology serves as a crutch to help the individual member of society through the emotional crises of life, such as the transition from childhood to adulthood, and the inevitability of death. It is important to realize, in this context, that science fiction has a large readership among the young, teenagers eager to find their places in society. And much of science fiction's tales of superheroes, time travel, and immortality are merely thinly-veiled attempts to deny the inevitability of death.

Without intending to, without even consciously realizing it until very recently, science fiction writers across the world have been slowly building up a mythology suitable for our modern scientific age. Like the scientists themselves, the writers share the rock-bottom faith in the power and beauty of rational human thought.

Contributors

HENRY H. BAUER is Dean of Arts and Sciences at Virginia Polytechnic Institute and State University. He has published books and articles on electrochemistry and analytical chemistry. He is currently interested in the societal role and impact of science, particularly as revealed in controversies over pseudoscience.

BEN BOVA, who served as the president of the National Space Society, is the author of more than seventy futuristic novels and nonfiction books dealing with science, space, and the future. Formerly the editor of *Analog* and *Omni* magazines, he has been active in the U.S. space program since its inception. A well-known lecturer and commentator on radio and TV, Mr. Bova holds a Master's degree in Communications from the State University of New York in Albany.

HONG-YEE CHIU is currently researching atmospheric physics problems at the NASA-Goddard Space Flight Center. He has worked in the fields of stellar evolution, neutrino processes, neutrino astronomy, supernovae, neutron stars, x-ray sources, intense magnetic fields, pulsars, and cosmology. He is the author of numerous books and articles on astronomy and is credited with coining the word "quasar."

S. V. M. CLUBE is Senior Research Fellow in Astrophysics, Department of Astrophysics at the University of Oxford. His special interest is terrestrial catastrophism and its relevance to the understanding of the solar system, the galaxy, and cosmology, as well as its historical implications. He has served with the Royal Observatories (Great Britain).

JOHN S. LEWIS is currently a Professor of Planetary Sciences at the University of Arizona. His previous position was Professor of Planetary Sciences at M.I.T. He holds a Ph.D. in physical and geological chemistry from the University of California in San Diego. He has worked principally on the application of chemistry to problems in the space sciences, especially the origin and evolution of the Solar System.

RAYMOND ARTHUR LYTTLETON is Professor Emeritus of Theoretical Astronomy at the University of Cambridge. He is a co-editor of, and contributor to *The Cambridge Encyclopaedia of Astronomy* (1977) and has published numerous other books and papers on astrophysics, cosmogony, cosmology, physics, and geophysics.

BRUCE SARGENT MACCABEE is Research Physicist at the Naval Surface Weapons Center in Maryland. He holds a Ph.D. in physics from American University. He has done extensive research in the fields of electronics, lasers, optics, and UFOs. He has written numerous books and articles on these subjects.

LLOYD MOTZ, Professor of Astronomy at Columbia University, has done extensive work on stellar interiors, nuclear physics, thermonuclear reactions, unified field theory, geometrical optics, and other related fields. He is currently engaged in a study of the structure of elementary particles using a gravitational model.

JAMES E. OBERG is a professional space operations specialist at NASA's Mission Control in Houston, Texas. He is also a widely published author on space topics, including the Soviet program on which he is an internationally recognized expert, and has authored eight books and several hundred magazine articles.

BRIAN O'LEARY received a Ph.D. in astronomy from the University of California in Berkeley in 1967 and was selected that year as an Apollo scientist-astronaut. His varied career includes faculty positions at Cornell, Cal Tech, Hampshire College, and Princeton; and he served as energy consultant to the U.S. Congress. Author and editor of several trade and technical books and articles, Dr. O'Leary is a leading researcher and spokesman for future human opportunities in space.

RAPHAEL PATAI is Professor Emeritus of Anthropology at Fairleigh Dickinson University in Rutherford, New Jersey. He received a Ph.D. from the Budapest University and from the Hebrew University of Jerusalem. He taught at the Hebrew University, and in 1944 founded the Palestine Institute of Folklore and Ethnology. He came to the United States in 1947 and taught anthropology at various universities. His publications are on the subjects of Israel, its sociology and ethnology, and the Middle East.

SHAUL SHAKED received his Ph.D. from the University of London and, since 1965, has taught at the Hebrew University in Jerusalem. His main publications include *A Tentative Bibliography of Geniza Documents*, an edition of *Wisdom of the Sassanian Sages*, and *Amulets and Magic Bonds: Aramaic Incantations of Late Antiquity* (with J. Naveh).

G. SETH SHOSTAK has started his own firm, Digital Images Computer Animation. He holds degrees in physics and astronomy and has pursued extragalactic radio astronomical research in Charlottesville, Virginia, and in the Netherlands. Since 1954 he has made films about astronomy that have won international awards.

S. FRED SINGER is Professor Emeritus of Environmental Sciences at the University of Virginia. He received degrees from Ohio State and Princeton University in engineering and physics. He served as Scientific Liaison Officer in Europe for the Office of Naval Research. He was appointed as the first Director of the U.S. Weather Satellite System by the Department of Commerce, later served as Deputy Assistant Secretary of Interior for water quality and research, and as Chief Scientist of the U.S. Department of Transportation. He has done research in astrogeophysics, and written and edited books dealing with population growth, global environment, and energy.

KURT STEHLING is the Senior Scientist and Chief Scientist Emeritus at the National Oceanic and Atmospheric Administration of the U.S. Department of Commerce. He is the author of six books and some two hundred technical papers and articles and has won numerous awards for his distinguished scientific accomplishments.

WILLIAM H. STIEBING, JR. has, since 1967, taught ancient history and archaeology at the University of New Orleans where he is now Professor of History. He received his B.A. degree in history from the University of New Orleans and a Ph.D. in Oriental Studies (Near Eastern Section) from the University of Pennsylvania. Dr. Stiebing specializes in ancient Near Eastern studies, particularly the history and archaeology of ancient Palestine. He has participated in archaeological excavations in Jordan and Lebanon.

CHARLES R. TOLBERT is currently an Associate Professor of Astronomy at the University of Virginia in Charlottesville, Virginia. He received his doctorate from Vanderbilt University and then served as Scientific Officer at the University of Groningen in Holland for four years. He has done research in both optical and radio astronomy and has been active in the fields of stellar classification and observational galactic structure, especially as related to variable stars and binary stars.

SEBASTIAN VON HOERNER is working as a consultant for various telescope projects in five countries. His fields of work have included astrophysics, turbulence, the N-body problem, stellar dynamics, gravitational collapse, and cosmology. He then changed to the field of engineering, concentrating on basic principles of telescope design and structural and optical optimization.

Index